Chemistry
for
University Students

Volume 1

Said A. H. Vuai B.Sc.,M.Sc., Ph.D.
Professor
Department of Chemistry,
School of Physical Sciences,
College of Natural &Mathematical Sciences,
The University of Dodoma.

and

H. C. Ananda Murthy B.Sc., M.Sc., M.Phil., Ph.D.
Assistant Professor
Department of Chemistry,
School of Physical Sciences,
College of Natural &Mathematical Sciences,
The University of Dodoma.

TANZANIA EDUCATIONAL PUBLISHERS LTD

Tanzania Educational Publishers Ltd,
TEPU House,
Uganda road, Plot No. 45, Block MDA,
Mob:0685 997583/0758 147871
Email: tepultd@yahoo.com
Website: www.tepu.co.tz
P.O. Box 1222,
Bukoba, Tanzania.

ISBN 978 9987 07 053 4

Dedication

We dedicate this book to our families

CONTENTS

Chapter 4

TRANSITION   ELEMENTS CHEMISTRY (*d-* BLOCK ELEMENTS)............... 89

Chapter 5

INNER TRANSITION ELEMENTS (*f-* BLOCK ELEMENTS)...................... 140

About the Authors

H. C. Ananda Murthy has been a sincere, committed and dedicated faculty member at various prestigious universities in India and Tanzania for the last 19 years. He is currently working as Assistant Professor, Department of Chemistry, The University of Dodoma, Tanzania, East Africa.

He completed his B.Sc. from Bangalore University, India in 1995. Later he completed his master's degree in Chemistry from Bangalore University, India in 1997. He also obtained M.Phil. degree in chemistry from Bharathidasan University, India in the year 2006. Then he went on to complete his Ph.D., for his work on corrosion behavior of aluminium composite materials from Visvesvaraya Technological University, Belgaum, India in the year 2012.

He has authored few books, compendia and published several research articles in the journals of international repute. He has taught General chemistry, Coordination chemistry and Organometallic Chemistry, Transition Elements Chemistry, Organometallic Chemistryand Electrochemistry & Corrosion protection, Advanced Coordination Chemistry, Surface Chemistry and Corrosion Science, Environmental Chemistry, Engineering chemistry and Advanced spectroscopyto the university students in India and Tanzania.

His research interest mainly includes Corrosion Science, Composite materials and Materials chemistry. He also presented many papers in national as well as international conferences. He has successfully completed a project titled "Amphibious bicycle" sanctioned by National Innovation Foundation of India (NIFI).

Said A. H. Vuai was born in Pemba, Zanzibar, Tanzania and grew up in Pemba. He received BSc. Degree in Chemical and Process Engineering from University of Dar es Salaam, Tanzania and his MSc. and PhD in Marine and Environmental Sciences majoring in Chemistry from University of the Ryukyus, Okinawa, Japan. After his PhD studies, he joined department of Sciences, of the State University of Zanzibar, where he taught Chemistry from 2004 to 2008. He then moved to the University of Dodoma in the Department of Physical Science where he continued to teach chemistry.In 2013-2014, he was awarded post Doctoral Fellowship to undertake Environmental Sustainability Assessment at the University of Siena, Italy.

Professor Vuai taught wide range of Chemistry courses, including Inorganic chemistry, organometalic chemistry, Analytical Chemistry, Environmental Chemistry and Material Sciences. He also served as external examiner for graduate students from Makerere University, Ugunday and Addis Ababa, University, Ethiopia.Professor Vuai has served number of administrative position at university of Dodoma, including Head of Department of Physical Sciences for three years, and Dean, School of Mathematical Sciences for about 5 years. He is enjoying wide collaborations in local regional and international institutions and researchers in Africa, Asia Europe and America. His research interest includes, Kinetic of chemical reactions, organometalic, Environmentalpollution of nutrients and trace metals and Public Health.

Acknowledgements

We express our profound gratitude and regards to the management of the University of Dodoma for providing us this opportunity to serve at CNMS, UDOM.

It is a great pleasure to express our gratitude to the Principal, CNMS for their constant support and encouragement in bringing out this book. We also place on record our appreciation to fellow colleagues in Department of Chemistry at CNMS, the Dean, School of Physical Sciences and UDOM for their cooperation.

INTRODUCTION

Chemistry for University Students is one book published in three volumes made up of chapters as follows:

Volume 1

Chapter 1: Electronic Structure of Atoms

Chapter 2: Periodic Table

Chapter 3: Chemical Bonding

Chapter 4: Transition Elements Chemistry

Chapter 5: Inner Transition Elements Chemistry

Volume 2

Chapter 6: Coordination Chemistry

Chapter 7: Electronic Spectra of Coordination Compounds

Chapter 8: Organometalic Chemistry

Volume 3

Chapter 9: Electrochemistry

Chapter 10: Environmental Chemitry

Chemistry has great contribution to the world. Chemistry provides thousands of raw materials to get finished products that help keep you safe, warm, cool, on time, in motion, and connected. From heart monitors to satellites to cell phones to microwave ovens to synthetic fibers, packed food items to medicine chemistry makes the things that make modern life possible.

Today's world would be inconceivable without the developments that chemistry and its applications have brought to science, engineering and technology. The achievements of chemistry are omnipresent. Students at universities definitely require chemical knowledge for their better understanding and academic performance. It is understood that a chemistry book comprising many aspects of chemistry is greatly needed by the students of the universities.

This book has been written keeping in mind the requirements of chemistry students at University level comprising of General chemistry, Periodic Table of elements, *d* and *f* block elements, Coordination chemistry, organometallic chemistry, Electrochemistry and Environmental Chemistry (All volumes).

We wish our book will be helpful to the students, as it is prepared to cater to the needs of the students, keeping in mind the relevance of each contents of the U G courses toward better understanding of the subject and with illustration including suitable figures, tables, and examples of structures and configurations of the complexes with simple presentation formats.

Chapter 1

ELECTRONIC STRUCTURE OF ATOMS

1.1 INTRODUCTION

Materials are made up of molecules and molecules contain small particles called atoms. Large number of discoveries, reported in the beginning of the 20th century helped to reveal the mystery of the atomic structure. The great minds like Max Planck, Albert Einstein, Rutherofrd, Bohr, Heisenberg, debroglie, Pauli, Schrodinger and many more enriched the scientific knowledge regarding atomic models and the quantum nature of the constituents of atoms as well as light. Let us understand the most prominent discoveries about the atomic nature of matter.

1.2 DALTON'S ATOMIC MODEL

John Dalton, in 1809 developed his famous theory of the atom. He regarded atom as a hard, dense and the smallest indivisible particle of matter. According to him, each element consists of particular kind of atoms. The properties of elements differ because of differences in the kind of atoms contained in them. Dalton's Atomic theory provided a satisfactory basis for the laws of chemical combination.

Certain experiments based on the passage of electricity through gases at low pressures revealed the presence of negatively and positively charged particle coming out of the gases. This experiment provides an evidence to convince that atom is not indivisible but consists of much smaller fundamental particles.

1.3 DISCOVERY OF ELECTRONS

Sir William Crookes showed in 1879 that when high potentials are applied to the metal plates in the highly evacuated discharge tube, so called cathode rays streamed from the negative pole, irrespective of the nature of the gas present in the discharge tube shown in Fig.1.

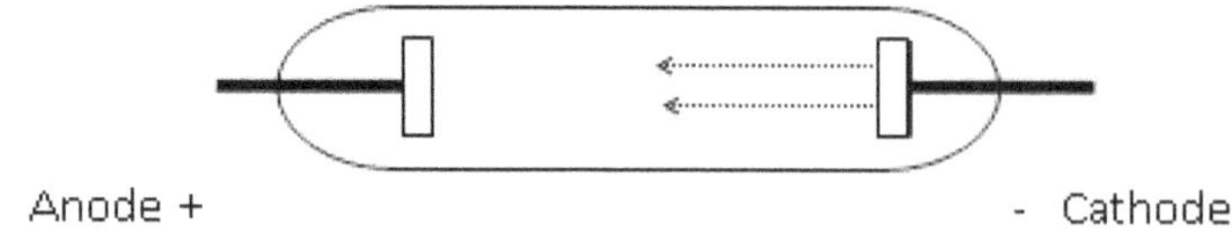

Fig. 1. Discharge tube

Cathode rays are the rays containing negatively charged particles which originate from cathode.
Important properties of cathode rays:
1. Cathode rays travel in straight lines from the negative electrode.
2. They produce fluorescence in the glass walls of the discharge tube.
3. They cast shadows if some target is placed in their path.
4. They can produce mechanical motion i.e, they cause light pedal wheel placed in their path to rotate.
5. They cause ionisation in gases.

6. They are deflected from their rectilinear path by electrostatic and magnetic fields.

These properties indicate that the cathode rays consist of negatively charged particles called electrons.

The negatively charged particles, which constitute the cathode rays, are called electrons.

The ratio of the charge to mass (e/m) of the cathode ray particles is observed to be same irrespective of the nature of the cathode material and nature of the gas.

Charge of an electron $\rightarrow$ -1.602x 10^{-19} coulomb

Mass of an electron $\rightarrow$ 9.1 x 10^{-31} kg

Positive rays (Canal rays)

In 1886 Goldstein used a discharge tube with a perforated cathode. He observed luminous rays passing backward through the perforation in the cathode. These produced flashes on the zinc sulphide plate placed at the end of discharge tube (Fig.2). These rays were show to consist of positively charged particles by Thomas and were called positive rays.

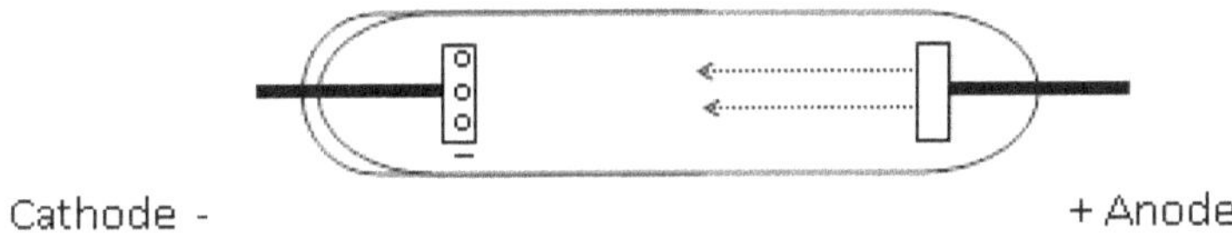

Fig. 2. Discharge tube

Obviously these positively charged particles are formed from the neutral atoms of the gas in the discharge tube by the loss of electrons.

The e/m ratio and the charge on these particles were found to vary in magnitude with the nature of the gas in the discharge tube. Their mass was found to be about the same as the atomic weight of the gas.

When the discharge tube contained hydrogen gas, the lightest positively charged particle was detected in the study of these positive rays.

$$H \rightarrow H^+ + e^-$$

Lightest particle called proton

Properties of positive rays

1. Positive rays are deflected in electric and magnetic fields. The direction of deflection shows that they are positively charged.

2. The value of e/m for these particles is very low compared with that of the electrons. The value of e/m depends on the nature of the gas in the discharge tube.

Proton is the particle obtained by the removal of an electron from a neutral hydrogen atom. The charge of proton is considered as unit positive charge. Charge of proton $\rightarrow$ +1.602 x 10^{-19} coulomb and Mass of proton $\rightarrow$ 1.6725 x 10^{-27} kg.

A proton is about 1837 times heavier an electron

$$\text{Mass of electron} = \frac{\text{Mass of proton}}{1837}$$

1.4 THOMSON'S MODEL OF ATOM

According to Thomson, the atom is a sphere of consisting of positively charged particle with number of electrons distributed within the sphere. This idea was soon discarded because this model could not explain the results of scattering experiment carried out by Lord Rutherford [1911].

1.5 DISCOVERY OF NUCLEUS: [Rutherford's α-particle scattering experiment]

In 1911 Rutherford performed a classic experiment for testing the Thomson's model. He bombarded thin filaments with high-speed α-particles [α-particles are doubly charged helium nuclei - $_2^4He^{2+}$]. Alpha particles are obtained from the radioactive element radium or polonium. The Direction in which a α-particle moved was detected with the help of a screen coated with zinc sulphide (Fig. 3).

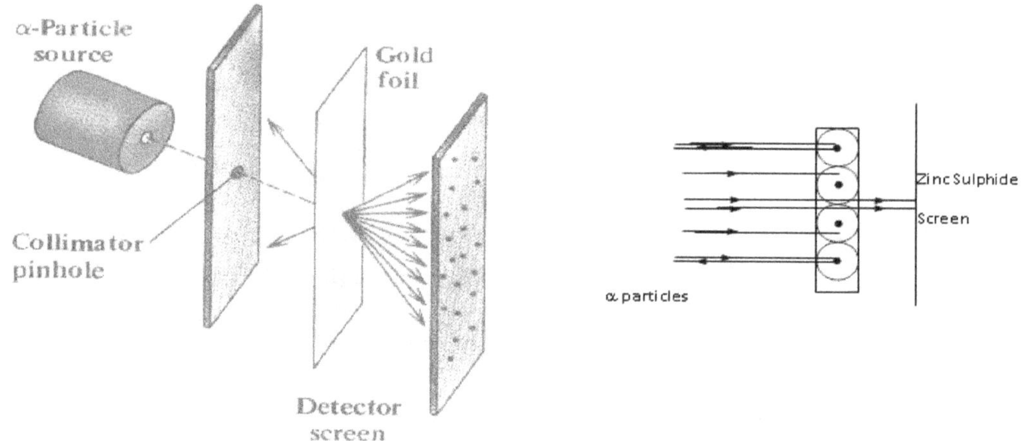

Fig. 3. Rutherford's alpha particle scattering Experiment

Rutherford interpreted experimental observations as follows:

1. Most of the alpha particles passed very nearly strait through the foil, passing of most of the alpha particles strait through it indicated that atom is extra-ordinary hollow with a lot of empty space inside.

2. A few α-particles were deflected from their original path through large angles, this suggested that the centre of the atom must be positively charged to repel positively charged α-particles. He named the positively charged central part as nucleus. The mass of the atom is concentrated in its nucleus.

3. A few α-particles were even turned back on their path which indicated that the nucleus is rigid from which α-particles could recoil.

1.6 RUTHREFORD'S ATOM MODEL: Assumptions

1. Matter is made up of atoms, which have both mass and volume. Atom consists of positively charged massive core at the centre called nucleus and electrons revolve around the nucleus in the large empty space.

2. Nucleus contains protons. Protons are positively charged particles. The electrons, which are present outside the nucleus, have the opposite electrical charge.

3. In a neutral atom, the number of electrons outside the nucleus is equal to the number of protons in the nucleus.

4. The volume of the nucleus is smaller than the volume of the atom by a ratio of about $1:10^{13}$.

5. To have a stable picture of the atom, Rutherford postulated that the electrons are revolving around the nucleus in closed orbits. Thus their centrifugal forces balance the force of attraction and keep them in their path.

Objection to Rutherford's Atom Model
According to Maxwell whenever an electric charge is subjected to acceleration, it emits radiation and loses energy. Bohr argued that if an electron moves round the nucleus in an orbit, it should be subjected to acceleration due to continuous change in its direction of motion. The electron should therefore, continuously emit radiation and lose energy. As a result its orbit should become smaller and smaller and finally it should drop into the nucleus. This however does not happen.

Discovery of Neutrons
A serious objection against the Rutherford model was as to how the electrons and protons were closely packed to give a stable nucleus. Rutherford solved this problem by suggesting that there might exist a particle carrying no charge but having a mass about the same as that of a proton. He named this particle as neutron.

It was in 1932 that Chadwick showed that neutrons are produced when α-rays impinge on certain light element such as Be or Boron.

Nuclei of atoms comprise of protons and neutrons, the only exception being hydrogen in which case nucleus is made up of only one proton.

Particle	Charge		Mass	Symbol
Electron	-1.602×10^{-19} C	-1	9.1085×10^{-31} kg	e [e$^-$]
Proton	$+1.602 \times 10^{-19}$C		1.67×10^{-27}kg 1.00728 amu	[p$^+$]
Neutron	0	+1	1.675×10^{-27}kg 1.008665 amu	[n^0]

1.7 BOHR'S ATOM MODEL

Neils Bohr a Danish physicist, put forward his model of atom in 1913. Bohr began with the assumption that electrons were orbiting the nucleus as much like the earth orbits the sun. We know from classical physics, a charge traveling in a circular path should lose energy by emitting electromagnetic radiation. If the "orbiting" electron loses energy, it should end up spiraling into the nucleus (which it does not). Therefore, classical physical laws either don't apply or are inadequate to explain the inner workings of the atom

Bohr borrowed the idea of quantized energy from Planck . He proposed that only orbits of certain radii, corresponding to defined energies, are "permitted" .An electron orbiting in one of these "allowed" orbits:

- ➢ Has a defined energy state
- ➢ Will not radiate energy
- ➢ Will not spiral into the nucleus

The important postulates of Bohr's theory are:

1. The electrons revolve around the nucleus in certain circular orbits called Stationary orbits. The energy of an electron remains constant as long as it [remains] stays in the same orbit. Each orbit is associated with a definite energy. These orbits are also called energy levels or energy shells.

2. Only those orbits are allowed for which the angular momentum of electron in these orbits should be an integral multiple of $\left(\dfrac{h}{2\pi}\right)$

i.e., Angular momentum of electron $\quad mVr = n.\left(\dfrac{h}{2\pi}\right)$

m → mass of the electron $\qquad$ v → velocity of the electron r $\rightarrow$ radius of an orbit   n $\rightarrow$ an integer (1,2,3,4, etc)

3. Energy is emitted or absorbed by an atom only when an electron moves from one level to another. Energy is emitted when an electron jumps down from a higher orbit to a lower one. When an electron absorbs energy, it jumps from a lower orbit to the higher orbit.

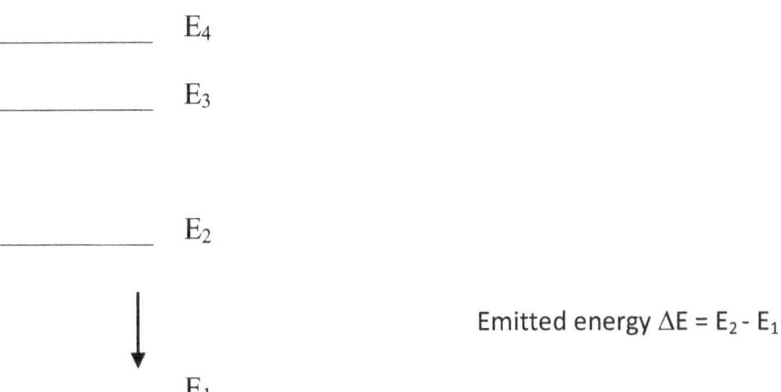

Energy is absorbed or emitted only in terms of small packets known as Quanta.

Limitations of Bohr's theory

1. Bohr's theory successfully explains the spectra of hydrogen atom and iso-electronic species like He^+, Li^{2+}, Be^{3+} but fails to explain the spectra of multi electron systems.

2. Bohr's theory fails to explain the fine structure of the spectrum.

3. It could not account for Zeeman effect [Splitting up of spectral lines in a strong magnetic field].

4. It could not account for stark effect. [Splitting up of spectral lines in a strong electric field]

5. It fails to recognise the wave property of electron, which was established in 1921 by

    de-broglie.

Mass Number: The sum of the number of protons and neutrons is called mass number. Since each proton and each neutron has a mass $\hat{n}\,1$, on the atomic weight scale, Atomic weight of an element is approximately equal to its mass number. Suppose, there is an atom of mass number A and atomic number Z, then, evidently, its nucleus consists of z protons and (A-Z) neutron. The number of electrons is equal to Z. For example, Lithium has mass no. (A) equal to 7 and atomic no.[z] equal to 3. Therefore its nucleus consists of 3 protons and (7-3) neutrons. There are 3 electrons outside the nucleus.

1.8 ELECTROMAGNETIC RADIATIONS

Electromagnetic radiations are the forms of radiant energy that travel as waves through space. Electromagnetic wave consists of two components. Electric and magnetic having vibrations in mutually ⊥r planes (Fig.4.).

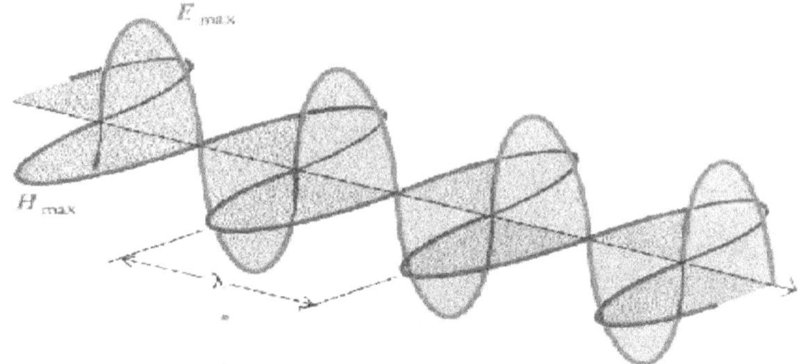

Fig.4. Electromagnetic radiations

Electromagnetic radiations (or) waves can be characterized by their wavelength, frequency and velocity which are interrelated quantities.

1. Wavelength [λ]: Wavelength is the distance between two neighbouring crests or troughs of the wave. It is expressed by two units

   i.   Armstrong $1A° = 10^{-10}$ m

   ii. Nanometer $1nm = 10^{-9}$ m

For example white light, which is considered to be mixture of different colours, contains waves between wavelength of 4000A° to 8000A° or 400 nm to 800 nm.

2. Frequency [γ] Frequency is the number of times a wave passes through a given point in one second. It is expressed in hertz. Hz [cycles per second].

3. Velocity [c]: Velocity is the distance traveled by a wave in one second. It is the velocity of light in air or volume and has a value 3×10^8 m/sec or 3×10^{10} m/sec.

Velocity C is the product of frequency and wavelength [C=γ x λ]

The reciprocal of the wavelength is called the wave number [$\bar{\gamma}$]

$$\bar{\gamma} = \frac{1}{\lambda} = \frac{\gamma}{c}$$

According to Planck's quantum theory, Light consists of packets of quanta of energy and the energy of each quantum of radiation depends up on the frequency of the radiation. The energy of the electromagnetic radiation is given by

$E = h\gamma$

where $h \rightarrow$ Planck's constant, $h = 6.626 \times 10^{-34}$ Js
Greater the frequency or smaller the wavelength of radiation, greater will be the energy of quanta.

Electromagnetic Spectrum

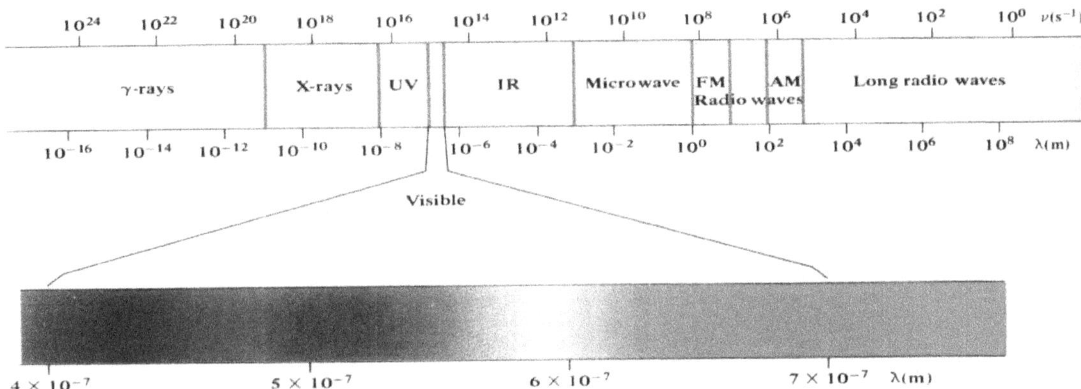

Spectrum:

When a radiation consisting of various frequencies is passed through a prism or grating gets itself resolved and the resultant radiation consists of several component frequencies. Such a pattern, which contains all the component frequencies are split up, is called a spectrum. A spectrum can be produced using an instrument called spectrometer.

Band Spectrum
An incandescent solid gives radiation, which produces a spectrum consisting of continuous bright bands of several colours. Such a spectrum is called Band spectrum.

Line Spectrum:
Incandescent gases or vapours emit radiations which when resolved gives a spectrum, which is discontinuous. It consists of bright separate spectrum is called line spectrum.

Electromagnetic spectrum:

Electromagnetic radiation	Frequency [Hz]
Radio Waves	10^{4} to 10^{9}
Microwaves	10^{9} to 10^{11}
Infrared	10^{12} to 10^{14}

Visible	4×10^{14} to 8×10^{14}
Ultra violet	8×10^{14} to 3×10^{18}
X-rays	10^{16} to 10^{19}
Gamma rays	$> 10^{19}$

1.9 LINE SPECTRUM OF HYDROGEN

When an electric discharge is passed through hydrogen gas at low pressure, the gas emits a rose coloured glow. The radiations pass out through the glass of the discharge tube. When a narrow beam of these radiations is passed through a prism, line spectrum consisting of several sharp lines is obtained. The spectrum is photo graphed (Fig. 5).

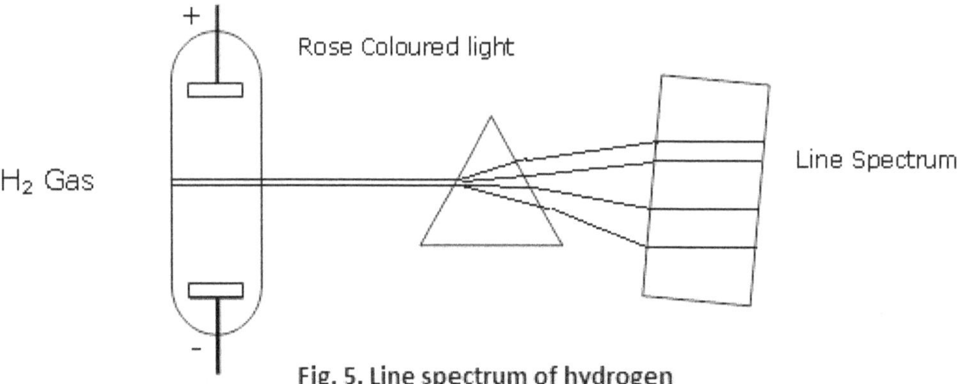

Fig. 5. Line spectrum of hydrogen

The spectral lines are found in the uv, visible and infra-red regions. Each line corresponds to a definite frequency.

The spectral lines can be grouped into several categories called Spectral series. There are five series and these are named after their discoverers.

Spectral series	Region in EMS
1. Lyman Series	Ultra violet region
2. Balmer series	Visible region
3. Paschen Series	Infra-red region
4. Brackettt series	Infra-red region
5. Pfund series	Infra-red region

The wavelength of any spectral line in any series is calculated using Rydberg's formula.

$$\bar{\gamma} = \frac{1}{\lambda} = R\left[\frac{1}{n_1^2} - \frac{1}{n_2^2}\right]$$

Where n_1 = 1.2.3.4.5 for the Lyman, Balmer, Paschen, Brackett and pfund series respectively and n_2 takes integral values starting from (n_1+1). R is a constant called Rydberg constant having a value 10.97×10^6 m^{-1}.

Balmer Series:

The Balmer series contains four prominent, bright, coloured lines. These are designated as $H_\alpha, H_\beta, G_\gamma$ and H_δ. The wavelength of any line in the Balmer series is given by the formula.

$$\frac{1}{\lambda} = R\left[\frac{1}{2^2} - \frac{1}{n^2}\right]$$

Where n can take up values 3,4,5,6,7,8,9 etc

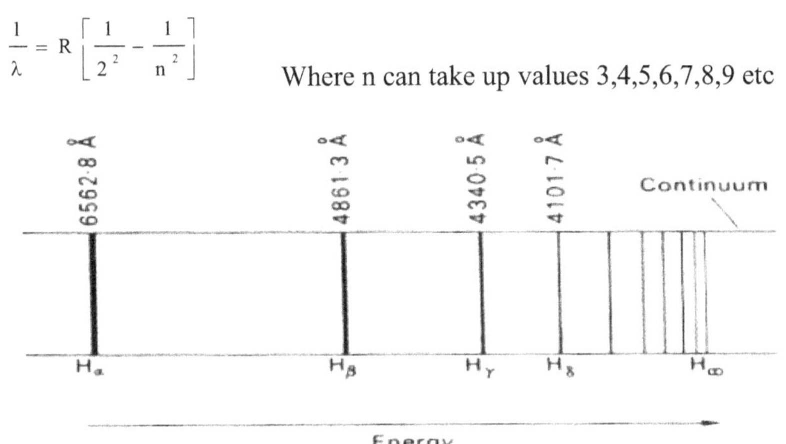

Spectrum of hydrogen in the visible region (Balmer series.)

Bohr's theory and Explanation of hydrogen spectrum

Neils Bohr explained the presence of various lines in the emission spectra of hydrogen atom based on his postulates.According to third postulate; Energy is emitted or absorbed by an atom only when an electron moves from one level to another. When hydrogen gas is subjected to electric discharge, energy is supplied. The molecules absorb energy and split up into atoms. The atoms also absorb energy. The hydrogen atoms get excited by allowing its only electron to jump into different higher energy levels. The electron in different atoms absorbs different quanta of energy and jump into different higher energy stages. But the atoms in the excited states (levels) are unstable. Therefore, the electrons jump back to different lower energy states in one or several steps. In each step, energy is emitted in the form of definite frequency. Therefore, though hydrogen atom has only one electron, the hydrogen spectrum consists of several lines, each corresponding to a definite frequency.

The jumping of electrons from various higher levels to the lowest level, that is to the k-shell (n=1, first orbit) gives the Lyman series of lines in uv region. Similarly the jump from different higher levels to L[n=2], M[n=3] N[n=4] and 0[4=5] shells cause the appearance of Balmer, paschen, Brackett and pfund series respectively.

The origin of hydrogen spectrum

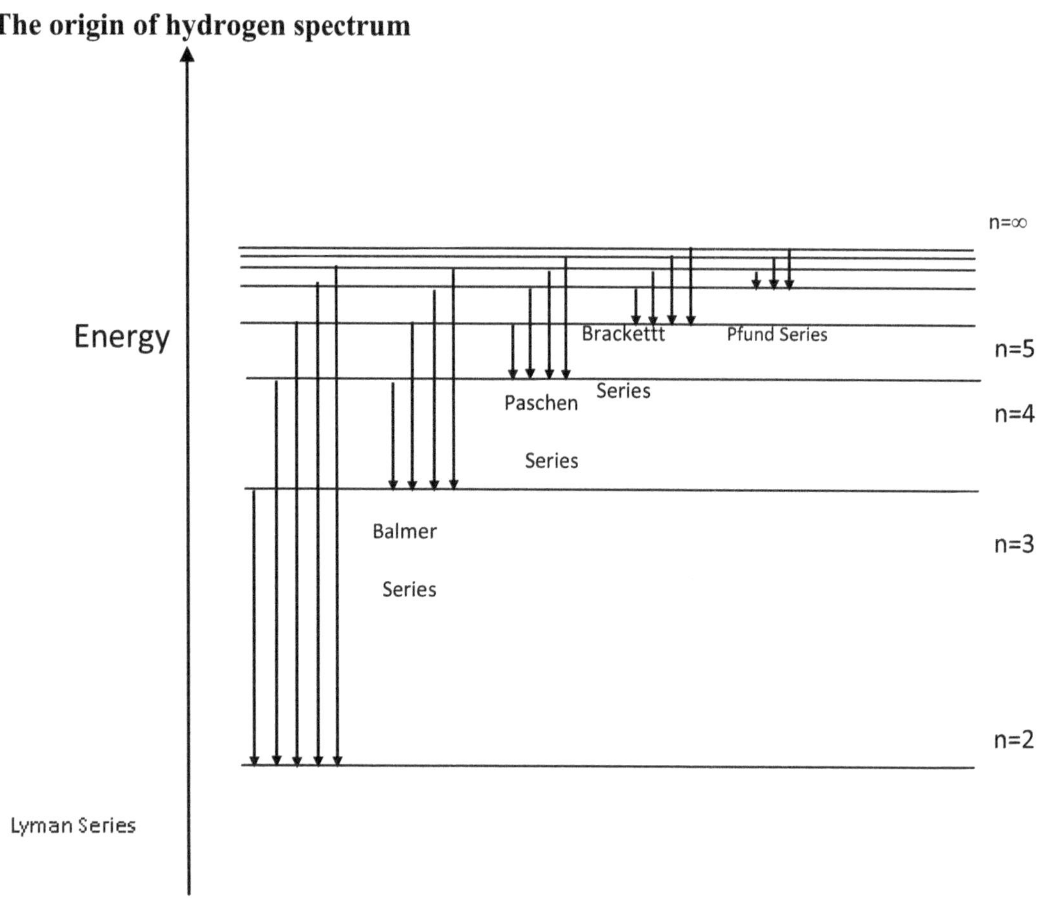

Spectral series	Values of n_1	Values of n_2	Spectral region	Wavelength in A^o
Lyman series	1	2,3,4,5….	Ultraiolet	<4000
Balmer series	2	3,4,5,6,…	Visible	4000-8000
Paschen series	3	4,5,6,7,….	Infrared	>7000
Brackettt series	4	5,6,7,8,…	Infrared	>7000
Pfund	5	6,7,8,9,…	Infrared	>7000

1.10 PARTICLE AND WAVE CHARACTER OF ELECTRON

Einstein had suggested in 1905 that light has a dual character as wave and also as particle. de Broglie proposed that matter also has dual characters, both as a material particle and as a wave. He derived an expression for calculating the wavelength λ of a particle of mass m moving with velocity v, according to which,

$$\lambda = \frac{h}{mv}$$

According to Einstein's equation, a photon of mass m is associated with energy

$$E = mc^2 ----\{1\}$$

According to Planck's equation,

$$E = h\gamma \qquad \text{but} \quad \gamma = \frac{c}{\lambda} \quad \text{therefore} \quad E = h\frac{c}{\lambda} \quad ----\{2\}$$

Equating   {1} and {2}

We get $\lambda = \dfrac{h}{mc}$ ----- de Broglie equation.

de broglie concept is significant only incase of small particles such as electron and is insignificant incase of large particle like Iron ball.

1.11 HEISENBERG'S UNCERTAINTY PRINCIPLE

Heisenberg (1926) proposed a principle, known as uncertainty principle with the help of which it is possible to understand the mutual relationship between particle and wave concept.Heisenberg uncertainty principle states that it is impossible to know exactly both the position and momentum of an electron or any other small particle simultaneously and accurately.

Thus when an electron behaves as a particle its position can be determined more accurately but there will be no certainty about its momentum. Similarly when an electron behaves as wave, the momentum can be determined accurately, but there will be no certainty about its position.

It Δx is uncertainty with regard to the position and Δp is uncertainty with regard to momentum, then uncertainty principle can be expressed as

$$\Delta x \times \Delta p \geq \frac{h}{4\pi}$$

In conclusion, the position of an electron moving with a definite velocity cannot be determined exactly. It is however possible to predict or state the probability of an electron is being at a given time.

The region in space around the nucleus where there is a maximum probability of finding an electron is called the atomic orbital.

Atomic orbitals are of different kinds and each orbital has a definite energy.

1.12 PARTICLE QUANTUM CHEMISTRY AND ATOMIC ORBITALS

Schrödinger's wave equation incorporates both wave- and particle-like behaviors for the electron. It opened a new way of thinking about sub-atomic particles, leading the area of study known as ***wave mechanics***, or ***quantum mechanics***.

Schrödinger's equation results in a series of so called wave functions, represented by the letter ψ*(psi)*. Although has no actual physical meaning, the value of ψ^2 describes the ***probability distribution of an electron***.

From Heisenberg's uncertainty principle, we cannot know both the location and velocity of an electron. Thus, Schrödinger's equation does not tell us the exact location of the electron, rather it describes the ***probability*** that an electron will be at a certain location in the atom.

Departure from the Bohr model of the atom

In the Bohr model, the electron is in a ***defined orbit***, in the Schrödinger model we can speak only of probability distributions for a given energy level of the electron. For example, an electron in the ground state in a Hydrogen atom would have a probability distribution which looks something like this (Fig. 6. A more intense color indicates a greater value for ψ^2, a higher probability of finding the electron in this region, and consequently, greater electron density):

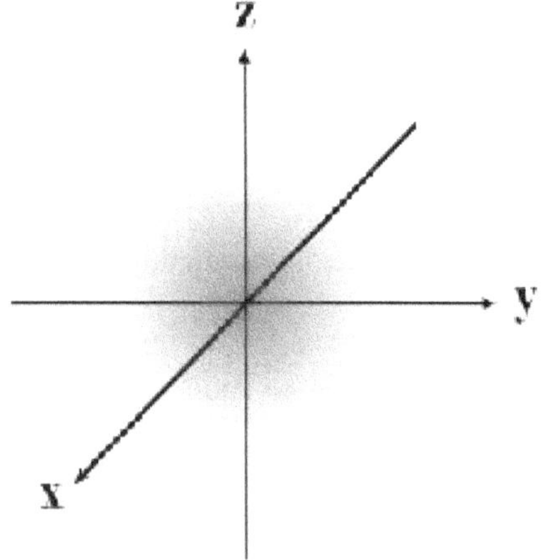

Fig. 6. *s*- orbital

Orbitals and quantum numbers

Solving Schrödinger's equation for the hydrogen atom results in a series of wave functions (electron probability distributions) and associated energy levels. These wave functions are called ***orbitals*** and have a characteristic energy and shape (distribution).

The lowest energy orbital of the hydrogen atom has an energy of -2.18 x 10^{18} J and the shape in the above figure. Note that in the **Bohr model** we had the same energy for the electron in the ground state, but that it was described as being in a defined *orbit*.

The Bohr model used a single quantum number (n) to describe an *orbit*, the Schrödinger model uses *three* quantum numbers: *n, l and m_l* to describe an *orbital*.

The principle quantum number 'n'
- Has integral values of 1, 2, 3, etc.
- As n increases the electron density is further away from the nucleus
- As n increases the electron has a higher energy and is less tightly bound to the nucleus

The azimuthal (second) quantum number 'l'
- Has integral values from 0 to (n-1) for each value of n
- Instead of being listed as a numerical value, typically 'l' is referred to by a letter (*'s'=0, 'p'=1, 'd'=2, 'f'=3*)
- Defines the **shape** of the orbital

The magnetic (third) quantum number 'm_l'
- Has integral values between 'l' and -'l', including 0
- Describes the orientation of the orbital in space

For example, the electron orbitals with a principle quantum number of 3 (i.e. *n*=3) would have the following available values of 'l' and 'm_l':

n *(principle quantum number)*	*l* *(azimuthal)* *(defines shape)*	*Subshell* *Designation*	*m_l* *(magnetic)* *(defines orientation)*	*Number of Orbitals in Subshell*
3	0	3s	0	1
	1	3p	-1,0,1	3
	2	3d	-2,-1,0,1,2	5

- A collection of orbitals with the same value of 'n' is called **an electron shell**
- A collection of orbitals with the same value of 'n' and 'l' belong to the same **subshell**

Thus:

- the **third electron shell** (i.e. 'n'=3) consists of the **3s, 3p** and **3d subshells** (each with a different shape)

- The *3s subshell* contains *1 orbital*, the *3p subshell contains 3 orbitals* and the *3d subshell* contains *5 orbitals*. (within each subshell, the different orbitals have different orientations in space)

- Thus, the *third electron shell* is comprised *of nine distinctly different orbitals*, although each orbital has the *same energy* (that associated with the third electron shell) *Note: remember, this is for hydrogen only.*

Restrictions on the possible values for the different quantum numbers (n, l and m_l) gives rise to the following patterns for the different shells:

- Each shell is divided into a number of *subshells equal to the principle quantum number* (e.g. the fourth shell is divided into four subshells: *s*, *p*, *d*, and *f*; whereas the first shell has a single subshell: *s*)

- Each subshell is divided into orbitals (*increasing by odd numbers*):

Subshell	Number of orbitals
s	1
p	3
d	5
f	7

The number and relative energies of all hydrogen electron orbitals through n=3 are shown below:

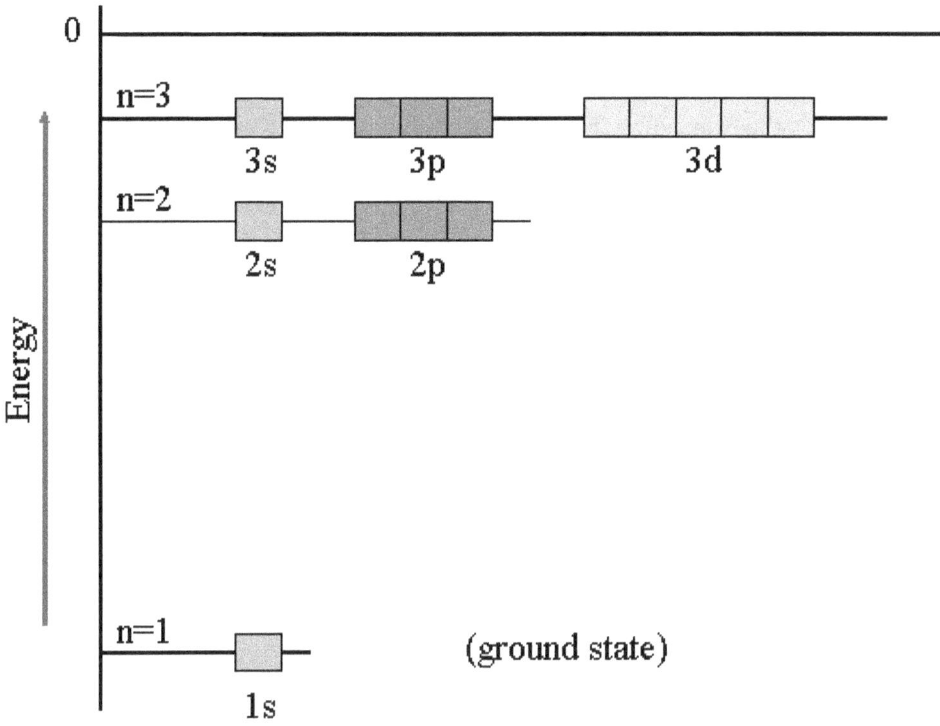

- At ordinary temperatures essentially all hydrogen atoms are in their ground states

- The electron may be promoted to an excited state by the absorbtion of a photon with appropriate quantum of energy

Representations of Orbitals

The *s* Orbitals

The *1s* orbital is spherically symmetrical. A plot of ψ^2 versus distance (r) from the nucleus shows a dramatic reduction in probability of finding the electron very far from the nucleus:

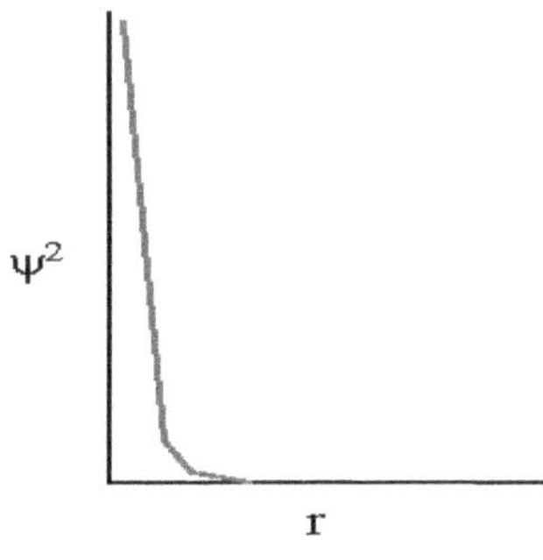

This indicates that in the ground state the electrostatic attraction of the electron for the proton in the nucleus is such that the electron is unlikely to be found far from the nucleus.

The higher energy s orbitals are also spherically symmetrical, however, they exhibit distinct nodes in the distribution probability:

In the higher s orbitals there exists *node* regions where the electron density approaches zero (2s has 1 node, 3s has 2 nodes, etc).

The higher s orbitals (excited states) have electron density distributions which indicate that there is a higher probability of finding the electron further away from the nucleus.

The size of the orbital increases as n increases

The most widely used representation of the Schrödinger orbits is to draw a

boundary which represents 90% of the total electron density distribution.

*For the s orbitals this would be a **sphere** representation.*

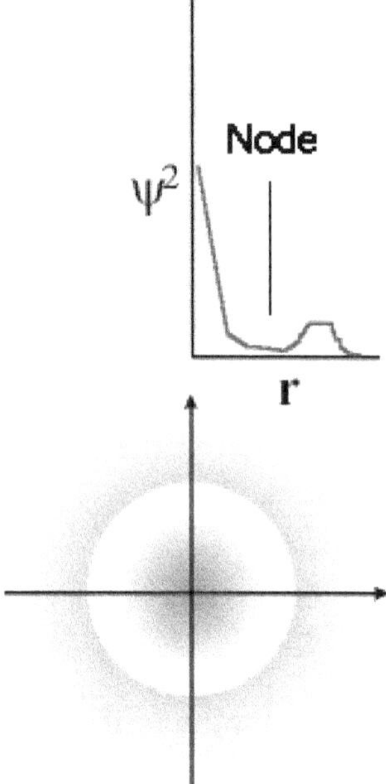

p Orbitals

- The ***p*** orbitals are 'dumbbell' shaped orbitals of electron density, with a node at the nucleus.

- There are three distinct p orbitals, they differ in their orientations

- There is no fixed correlation between the three orientations and the three magnetic quantum numbers (m_l)

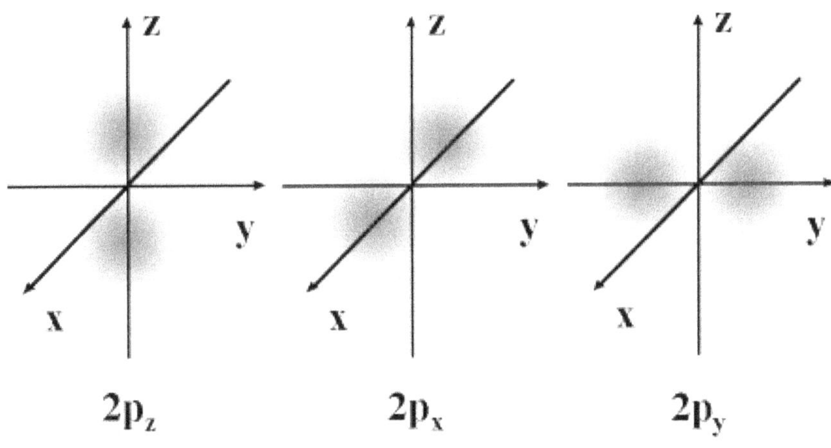

2p_z   **2p_x**   **2p_y**

Fig. 7. *p* - orbitals

The d and f orbitals

In the third shell and beyond there are five d orbitals, each has a different orientation in space:

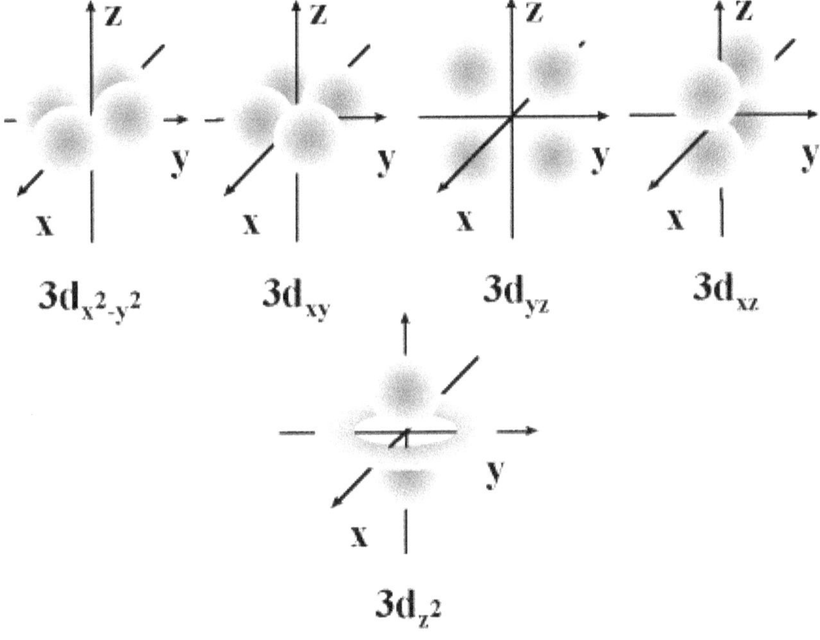

3d_{x²-y²}   **3d_{xy}**   **3d_{yz}**   **3d_{xz}**

3d_{z²}

Fig. 8. *d*- orbitals

Although the $3d_{z^2}$ orbital looks different, it has the same energy as the other *d* orbitals.

There are 7 equivalent *f* orbitals (for each value of ***n*** 4 or greater). They are pretty difficult to represent on a 3-d contour diagram.

Understanding orbital shapes is key to understanding the molecules formed by combining atoms

Orbitals in many-electron atoms

The hydrogen atom is a simple system having only *one electron*.

The quantum mechanical description of the hydrogen atoms places all subshells (i.e. *l* quantum number, or the *s*, *p*, *d* and *f* subshells) with the same principle quantum number (*n*) on the same energetic level.

An atom with more than 1 electron is called a *many-electron* atom.

Although the *shape* of electronic orbitals for many-electron atoms are the *same* as those for the hydrogen atom, *the presence of more than 1 electron influences the energy levels of the orbitals (due to electron-electron replusion)*.

For example, the 2s orbital is a lower energy state than the 2p orbital in a many-electron atom: (note: this is a qualitative representation for an "average" many-electron atom)

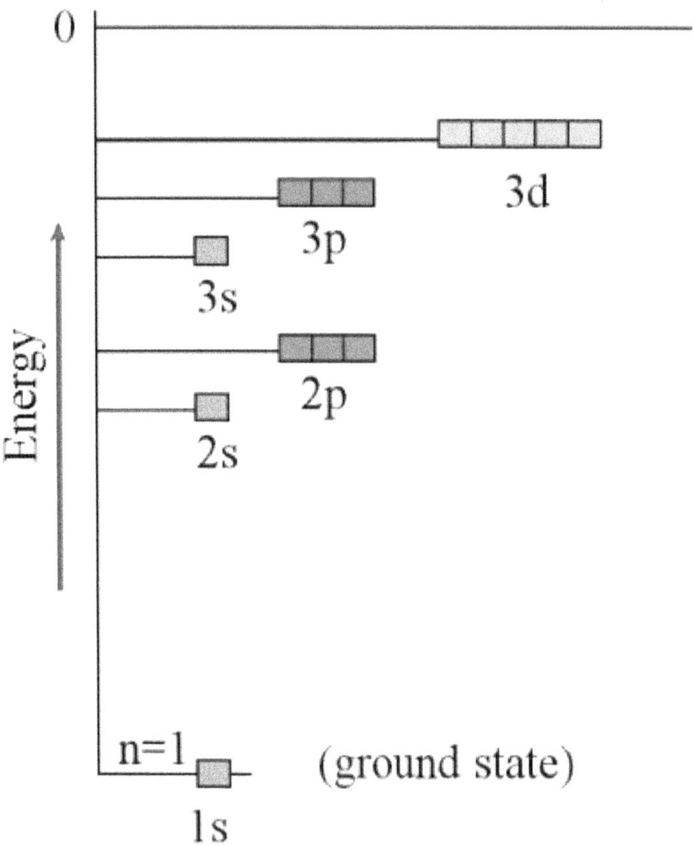

1.13 RULES FOR ELECTRONIC CONFIGURATIONS

The best representation of electronic configuration for elements is achieved through learning few important rules developed by Wolf-gang Pauli and Friedrich Hund.

a. *Pauli's Exclusion principle*

Pauli's exclusion principle states that "It is impossible for any two electrons in the same atom to have all the four quantum numbers identical"

This means each electron in an atom has only one set of values n, l, m_l and s. Any two electrons in the same orbital will have different values for at least one quantum numbers. An atomic orbital can accommodate only two electrons with opposite spins.

For example   4s-orbital   4s
n=4 l=0 m=0 s=+½   - clockwise electron
n=4 l=0 m=0 s=-½ - Anticlockwise electron
              Same values   Different s values

The Pauli's exclusion principle limits the number of electrons that can beaccommodated in principal shells and sub shells (orbitals). These are illustrated in the table given below,

Main shell	Principal quantum number [n]	Azimuthal q.no [l] [values between 0 and(n-1)]	Subshell designation	Magnetic q. no.[m_l] (2l+1) no. of values. These values vary between –1 to +l.	No. of orbitals in sub shells	Principal shell capacity [$2n^2$]
K	1	0	1s	0	1	2
L	2	0	2s	0	1	2+
		1	2p	-1,0,+1	3	6 = 8
M	3	0	3s	0	1	2+
		1	3p	-1,0,+1	3	6+
		2	3d	-2,-1,0,+1,+2	5	10=18
N	4	0	4s	0	1	2+
		1	4p	-1,0,+1	3	6+
		2	4d	-2,-1,0,+1,+2	5	10+
		3	4f	-3,-2,-1,0,+1,+2,+3	7	14=32

b. Hund's rule of maximum multiplicity

According to this rule electron pairing in any of the s, p, d or f orbital is not possible until all the available orbitals of a given set contain one electron each.

For example, electronic configuration of N is $1s^2\ 2s^2\ 2p^3$. This indicates that 1s and 2s orbitals are completely filled with two electrons each. But according to Hund's rule, electrons occupy all the available orbitals singly with parallel spins, before they pair up in any orbital. Thus the three electrons of the 2p-orbital will have the configuration $2p_x^1, 2p_y^1, 2p_z^1$. Thus the new electronic configuration of nitrogen is $1s^2\ 2s^2\ 2p_x^1\ 2p_y^1\ 2p_z^1$

In case of oxygen 2p orbital has four electrons, first three electrons in p_x, p_y and p_z orbitals singly, fourth electron occupies p_x orbital and pairs with the electron already present in it. Thus the electronic configuration of oxygen is $1s^2 \, 2s^2 2p_x^2 \, 2p_y^1 \, 2p_z^1$

Hunds rule also states that the total energy of a many electron atom with more than one electron occupying a set of degenerate orbitals is lowest it, as far as possible, electrons occupy different atomic orbitals and have parallel spins. The lowest energy configuration for an atom is the one with maximum number of unpaired electrons.

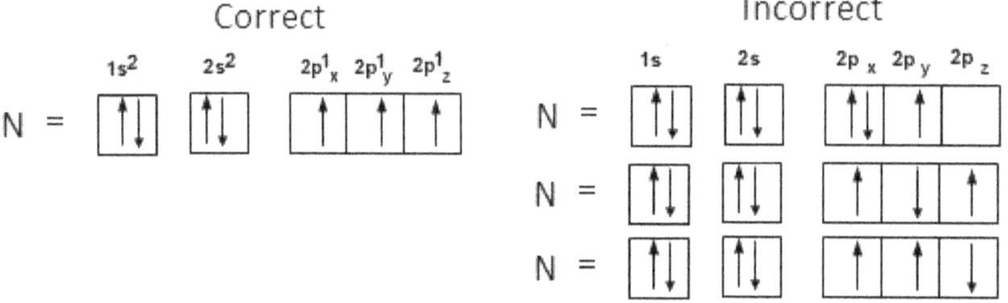

d. **Aufbau principle**

c. <u>(n+l) rule</u>
In neutral isolated atoms, the most stable orbitals is the one for which the sum (n+l) is least. When two orbitals have the same (n+l) value the orbital with lower value of n is the stabler one.

For example   for 4s sublevel $n = 4$ $l = 0$ $n + l = 4$

for 3d sublevel $n = 3$ $l = 2$ $n + l = 5$

For 4s sublevel (4+l) has lower value than for the 3d sublevel. So 4s level is occupied by electrons before the 3d level.

d. <u>Aufbau principle</u>

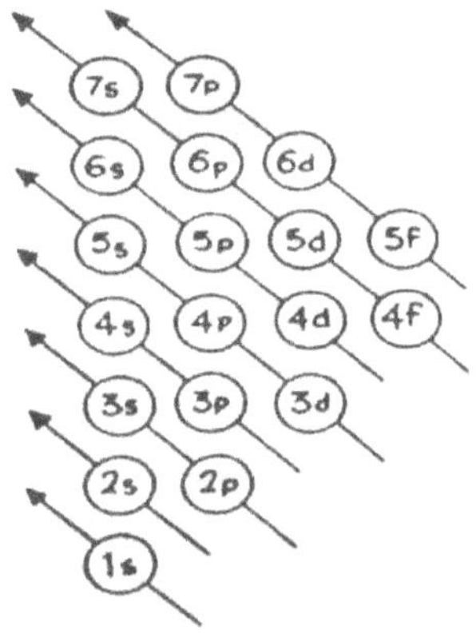

Energy level diagram

Aufbau principle states that the electrons enter the various orbital in the order of increasing energy. Aufbau principle describes the method which involves building up of electronic configuration for atom by adding on electron to the outer shell and one proton to the nucleus of that atom. Aufbau means building up in German language. Pauli's exclusion principle, Hund's rule, (n+l) rule and Afbau principle are employed in building the electronic configuration of atoms. The ascending order of energy of the sublevels, in accordance with aufbau principle is 1s, 2s,2p, 3s, 3p, 4s, 3d, 4p, 5s, 4d, 5p, 6s,4p, 5d, 6p, 7s, 5p, 6d, 7p

Many of the principles or rules discussed earlier, namely Afbau principle, Hunds rule, (n+l) ruleand Pauli's exclusion principle provide us enough information in filling of electrons to various orbitals of different elements. The first element, hydrogen atom (atomic number z = 1) having only one electron in s orbital, is half filled i.e., the configuration is $\mathbf{1s^1}$. The electronic configuration is $\mathbf{1s^2}$ for the next element helium, He (z = 2). Maximum capacity of s-orbital is 2 in accordance with the formula $2n^2$ (where n-principal quantum no.) Now k-shell is completely filled. Next the third electron in Lithium enters 2s orbital of L-shell. The configuration of Li (z = 3) is **$1s^2\ 2s^1$**. In beryllium (z = 4), 2s orbital is completely filled, so the configuration of Be is **$1s^2\ 2s^2$**. Boron, the next element has five electrons of which 5th electron occupies 2p orbital of L-shell so the configuration of B (z =5) is **$1s^2\ 2s^2\ 2p_x^1$**. In carbon, 6th electron, does not get paired up with electron of $2p_x$ orbital but enters 2py [orbital] or subshell in accordance with Hund's rule. So the configuration of C(z = 6) is **$1s^2\ 2s^2\ 2p_x^1 2p_y^1$**. Nitrogen has electronicnic configuration with three of its electron in p three 2p orbitals **$1s^2\ 2s^2\ 2p_x^1 2p_y^1 2p_z^1$**. In oxygen, O (z = 8) the 8th electron gets paired with **$2p_x^1$** orbital and has the configuration **$1s^2\ 2s^2\ 2p_x^2 2p_y^1 2p_z^1$**. Similarly in fluorine and neon, 9th and 10th electrons pair up with **$2p_y^1$ and $2p_z^1$** orbitals respectively. Hence the configurations of fluorine (z =9) is **$1s^2\ 2s^2\ 2p_x^2 2p_y^2 2p_z^1$** and that of neon (z =10) **is $1s^2\ 2s^2\ 2p_x^2 2p_y^2 2p_z^2$**. The 2nd shell (L-Shell) is completed in neon.

The electronic configuration of the first 10 elements

At No.	Element	Symbol	Electronic configuration Spdf-notation	Orbital diagram (1s 2s 2p)
1	Hydrogen	H	$1s^1$	1s: ↑
2	Helium	He	$1s^2$	1s: ↑↓
3	Lithium	Li	$1s^2\ 2s^1$	1s: ↑↓ 2s: ↑
4	Beryllium	Be	$1s^2\ 2s^2$	1s: ↑↓ 2s: ↑↓
5	Boron	B	$1s^2\ 2s^2\ 2p_x^1$	1s: ↑↓ 2s: ↑↓ 2p: ↑ □ □
6	Carbon	C	$1s^2\ 2s^2\ 2p_x^1 2p_y^1$	1s: ↑↓ 2s: ↑↓ 2p: ↑ ↑ □
7	Nitrogen	N	$1s^2\ 2s^2\ 2p_x^1 2p_y^1 2p_z^1$	1s: ↑↓ 2s: ↑↓ 2p: ↑ ↑ ↑
8	Oxygen	O	$1s^2\ 2s^2\ 2p_x^2 2p_y^1 2p_z^1$	1s: ↑↓ 2s: ↑↓ 2p: ↑↓ ↑ ↑
9	Fluorine	F	$1s^2\ 2s^2\ 2p_x^2 2p_y^2 2p_z^1$	1s: ↑↓ 2s: ↑↓ 2p: ↑↓ ↑↓ ↑
10	Neon	Ne	$1s^2\ 2s^2\ 2p_x^2 2p_y^2 2p_z^2$	1s: ↑↓ 2s: ↑↓ 2p: ↑↓ ↑↓ ↑↓

The third shell -M-shell gets its quota of electrons from sodium to Argon. The noble gas core abbreviated electron configuration for the elements is given below:

11	Sodium	Na	$1s^2\,2s^2\,2p^6\,3s^1$ or [Ne] $3s^1$
12	Magnesium	Mg	[Ne] $3s^2$
13	Aluminium	Al	[Ne] $3s^2 3p^1$
14	Silicon	Si	[Ne] $3s^2 3p^2$
15	Phosphorous	P	[Ne] $3s^2 3p^3$
16	Sulphur	S	[Ne] $3s^2 3p^4$
17	Chlorine	Cl	[Ne] $3s^2 3p^5$
18	Argon	Ar	[Ne] $3s^2 3p^6$

The fourth shell, electron filling starts from K, where 19th electron moves to 4s orbital rather to 3rd orbital in accordance with (n+l) rule.

19	Potassium	K	$1s^2\,2s^2\,2p^6\,3s^2\,3p^6\,4s^1$ or [Ar] $4s^1$
20	Calcium	Ca	$1s^2\,2s^2\,2p^6\,3s^2\,3p^6\,4s^2$ or [Ar] $4s^2$
21	Scandium	Sc	[Ar] $4s^2 3d^1$
22	Titanium	Ti	[Ar] $4s^2 3d^2$
23	Vanadium	V	[Ar] $4s^2 3d^3$
24	Chromium*	Cr	[Ar] $4s^1 3d^5$
25	Manganese	Mn	[Ar] $4s^2 3d^5$
26	Iron	Fe	[Ar] $4s^2 3d^6$
27	Cobalt	Co	[Ar] $4s^1 3d^7$
28	Nickel	Ni	[Ar] $4s^2 3d^8$
29	Copper*	Cu	[Ar] $4s^1 3d^{10}$
30	Zinc	Zn	[Ar] $4s^2 3d^{10}$

Both chromium and Copper have unusual electronic configuration deviating form normal electron filling process because of a general rule, which states that "completely filled and completely half filled configurations are more stable than other configuration". In the absence of this rule Cr and Cu should have the following configuration respectively, [Ar] $4s^2 3d^4$ and [Ar] $4s^2 3d^1$.

31	Gallium	Ga	[Ar] $4s^2 3d^{10} 4p^1$
32	Germanium	Ge	[Ar] $4s^2 3d^{10} 4p^2$
33	Arsenic	As	[Ar] $4s^1 3d^{10} 4p^3$
34	Selenium	Se	[Ar] $4s^2 3d^{10} 4p^4$
35	Bromine	Br	[Ar] $4s^1 3d^{10} 4p^5$

36	Krypton	Kr	[Ar] $4s^2 3d^{10}4p^6$
37	Rubidium	Rb	[Kr] $5s^1$
38	Strontium	Sr	[Kr] $5s^2$
39	Yttrium	Y	[Kr] $5s^2 4d^1$
40	Zirconium	Zr	[Kr] $5s^2 4d^2$
41	Niobium*	Nb	[Kr] $5s^1 4d^4$
42	Molybdenum*	Mo	[Kr] $5s^1 4d^5$
43	Technetium	Tc	[Kr] $5s^2 4d^5$
44	Ruthenium*	Ru	[Kr] $5s^1 4d^7$
45	Rhodium*	Rh	[Kr] $5s^1 4d^8$
46	Palladium*	Pd	[Kr] $5s^0 4d^{10}$
47	Silver *	Ag	[Kr] $5s^1 4d^{10}$
48	Cadmium	Cd	[Kr] $5s^2 4d^{10}$
49	Indium	In	[Kr] $5s^2 4d^{10}5p^1$
50	Tin	Sn	[Kr] $5s^2 4d^{10}5p^2$
51	Antimony	Sb	[Kr] $5s^2 4d^{10}5p^3$
52	Tellurium	Te	[Kr] $5s^2 4d^{10}5p^4$
53	Iodine	I	[Kr] $5s^2 4d^{10}5p^5$
54	Xenon	Xe	[Kr] $5s^2 4d^{10}5p^6$

1.14 EXCERCISE QUESTIONS

1. Calculate the amount of energy associated with the following radiations
Red radiations of wavelength $7000A^o$
Violet radiations of wavelength of $4000A^o$.
2. Calculate the wavelength of a wave of frequency 1014 Hz and traveling with a speed of
3 x 108 m/s
3. The wavelength of a wave is 620 nm. Calculate the frequency and wave no.
4. The wave no. of a wave is 3.3 x 105 /m. Find the wavelength frequency and energy of
that wave.
5. Calculate the no. of photons of light with a wavelength of $6000A^o$ that provide 1 joule of energy
$E = h\gamma$ make 1 photon.
6. Find out the frequency of the limiting line in the Balmer series.
7. The first line in the Balmer series of hydrogen spectrum has the wavelength 656 /nm.
Find the energy decrease as this photon is emitted.
8. A microscope using suitable photon is employed to locate on electron in an atom with in   a
distance of $0.1A^o$. What is the uncertainty involved in the measurement of its velocity?
$m = 9.1 \times 10^{-31}$ kg $k = 6.626 \times 10^{-34}$ JS
9. A cricket ball weighing 100g is to be located with 0.01 A^o. What is the uncertainty in its
velocity? Comment on your result.
10. Calculate the debroglie wavelength associated with a particle moving with a velocity
9.25 x 107 m/s, h=6.62 x 10^{-34} JS mass of an electron=9.1 x 10^{-31} kg.

Exercise Questions: Type 1

1. Dalton considered atom as a ________
2. What are cathode rays?
3. What are anode rays?
4. The ratio e/m remains constant for cathode rays, give reason.
5. What is the charge of proton and electron?
6. Give the ratio of mass of electron to the mass of proton.
7. What are α-particles?
8. Which experiment led to the discovery of nucleus?
9. What are Nucleons?
10. According to Bohr, the angular momentum of electron mvr is equal to ____
11. Define mass number and atomic number
12. What are electromagnetic radiations?
13. Give an expression relating frequency, velocity and wavelength of radiations.
14. Give the expression relating energy and frequency of the radiation.
15. Define frequency and wavelength of EM radiations
16. What do you mean by spectrum?
17. Which expression is used to find the wavelength of spectral lines of hydrogen spectrum?
18. What is the value of Rydberg constant?
19. Write Rydberg's formula for the Balmer series of hydrogen spectrum.
20. Which part of the EMS region containing Balmer and Paschen series spectral lines?
21. What is the frequency of radiation with a wavelength 412 nm?
22. Give de-broglie equation
23. Give the statement of Heisenberg's uncertainty principle.
24. Give the mathematical expression of HUP
25. Define an atomic orbital.
26. Write the shapes of p-orbitals
27. What are quantum numbers?
28. Name the four-quantum numbers?
29. How many values that Azimuthal quantum number l can have for a given value of principle quantum number n?
30. What is the value of l for 3d orbital?
31. How many values that m can have for a given value of l? Name them
32. What is the maximum capacity of L-shell
33. State Pauli's exclusion principle
34. What is Hund's rule of maximum multiplicity?
35. What is (n+l) rule?
36. What is Afbau principle?
37. Write the electronic configuration for the following metals
    i. Chromium   ii. Ion   iii. Copper
38. Which of the following orbitals is most stable?
    a) 4p   b) 3d

Exercise Questions: Type 2

1. What are the Isotopes? Give examples.
2. What are the merits of Bohr's theory?
3. Write Rydberg formulae and explain the terms.
4. Write debroglie equation and explain the terms.
5. State Heisenberg's uncertainty principle.
6. Which are the four quantum numbers to represent an electron?
7. What are the possible values of l and m when $n = 4$?
8. State and explain Hund's rule of maximum multiplicity.
9. State Aufbau principle.
10. Nitrogen is chemically unreactive. Give reason.
11. Number of unpaired electrons present in
    a) Cu^{2+} b) Mn^{2+}
12. State and explain Pauli's exclusion principle.

Exercise Questions: Type 3

1. What are the postulates of Bohr's theory?
2. Explain the origin of hydrogen spectrum using Bohr's theory.
3. What are quantum numbers? Explain the significance of all the quantum numbers.
4. What are atomic orbitals? Sketch the shape of s,p,d and f orbitals.
5. Explain the sequence in which electrons are filled into an atom in the ground state.
6. What is (n+l) rule? Give two examples.

Objective Type Questions

1. Which of the following statements about cathode rays is wrong?
   a. Cathode rays produce a mechanical effect.
   b. The ratio e/m is independent of the nature of gas
   c. The ratio e/m of cathode rays is considerably smaller than for positive rays.
   d. The rays carry a negative charge.
2. Which of the following statements contradicts Daltons Atomic theory?
   a. A chemical change involves rearrangement of atoms
   b. Atoms form ions by gaining or losing electrons
   c. Atomic is an ultimate particle of matter
   d. The total number of atoms during a chemical change remains unaltered.
3. What is wrong about anode rays?
   a. The e/m ratio is constant
   b. They are deflected by electrical and magnetic field
   c. They are produced by ionization of molecules of the residual gas.
   d. They are positively charged particles
4. In α-ray scattering experiments, one in 10,000 was deflected back. Most of the α-particles, which were not deflected back.
   a. Caused the gold foil to become radio active
   b. Were absorbed by gold foil
   c. Passed through extra nuclear part of the atom
   d. Were neutralized by the negative electrons.
5. An atom with atomic number 92 and mass number 238 contains.

a. 92 protons
b. 92 protons, 92 electrons and 128 neutrons
c. 92 electrons, 92 neutrons
d. All are wrong

6. An atom has net charge of -1. It has 18 electrons and 20 neutrons. It's mass number is
    a. 37   b. 35   c. 38   d. 20

7. The fundamental particle which is responsible for keeping the nucleus together is
    a. Proton   b. neutron   c. Positron   d. Meson
    [Exchange of mesons between protons and neutrons holds the nucleus together]

8. Bohr's model can explain
    a. The spectrum of hydrogen atom only
    b. The spectrum of atom or ion containing one electron only
    c. The spectrum of hydrogen molecule
    d. The solar spectrum

9. Bohr's atomic model added to the Rutherford atom the restriction that
    a. Light is emitted by orbiting electron
    b. All nuclei of a given element have the same mass
    c. All electrons have the same mass
    d. Only certain electron orbits are allowed

10. Which of the following statements contradicts Bohr's model of hydrogen atom.
    a. Energy of an electron in the orbit is quantized
    b. Electrons revolve around the nucleus in discrete circular orbits
    c. Orbiting electron should radiate energy
    d. Energy is emitted from a atom when in jumps from high energy level to low level.

11. Which property is common to all types of electromagnetic radiations?
    a. Frequency   b. Velocity   c. Wavelength   d. Wave number

12. Which of the following electromagnetic radiation has extremely small wavelength?
    a. Radio wave b. cosmic rays   c. IR rays   d. Microwaves

13. The ratio of energy of photon of wavelength 3000 A$^{\circ}$ and 600A$^{\circ}$ is
    a. ½   b. 1/3   c. 2   d. 1/6

14. In hydrogen spectrum most energetic transition of electrons are found in the following series.
    a. Balmer   b. Lyman   c. Brackettt   d. pfund

15. The line spectrum of two elements is not identical
    a. They do not have same number of neutrons
    b. They have dissimilar mass number
    c. They have different energy level schemes
    d. They have different number of electrons

16. Which of the following transitions of electrons in the hydrogen atom emits maximum energy
    a. $n_2 \rightarrow n_1$ b. $n_1 \rightarrow n_4$ c. $n_4 \rightarrow n_3$ d. $n_3 \rightarrow n_1$

17. Bohr theory is not applicable in case of
    a. H   b. H$^+$ c. He^{2+} d. Li^{2+}

18. The uncertainty in the position of the electron [mass 9.1×10^{-28}g] moving with a

velocity of 3 x 104 m/s accurate up to 0.011% will be.

19. The limiting line in the Balmer series will have a frequency of
   a. $3.29 \times 1015\ s^{-1}$ b. $-3.65 \times 1014\ s^{-1}$
   c. $8.22 \times 1014\ s^{-1}$ d. $-8.21 \times 1014\ s^{-1}$
   for limiting the $n_1=2\ n_2=10$

20. Momentum of particle having 10^{-17} m debroglie wavelength is
   a. $13.25 \times 10^{-17}\ kg\ ms^{-1}$
   b. 26.5×10^{-7}
   c. 6.625×10^{7}
   d. 3.3125×10^{-7}

21. The maximum number of $3d$ electrons having spin quantum number   is
   a.   10   b. 14   c. 5   d. Any number from 1 to 10

22. d-orbital can accommodate a maximum of
   a.   7 electrons b. 10 electrons   c. 5 electrons   d. 6 electrons

23. How many electrons can fit into the orbitals that comprise the 3rd quantum shell n=3.
   a.   10   b. 18   c. 20   d. 32

24. How many electrons can be accommodated in an orbital with l=3? Among them how many electrons can have s=+½?
   a. 7,7   b. 14,7   c. 14,14   d. 7,14

25. Principal, azimuthal, and magnetic quantum numbers are respectively related to?
   a.   Shape, size and orientation of orbital
   b.   Size, shape and orientation of orbital
   c.   Size, orientation and shape of orbital
   d.   None

26. Which of the following sets of quantum number is not allowed?
   a. n=2, l=1, m=0   b. n=2, l=2, m=-1
   c. n=3, l=0 m=0   d. n=3, l=1, m=-1

27. What is the correct orbital designation of an electron with the quantum number n=4, l=3 m=-2 s=½
   a. 35   b. 47   c. 7p   d. 65

28. Which set of quantum numbers is not correct in a 3d orbital?
   a.3,2,-2, +½   b.3,-2,0,-½
   c.3,4,-1, +½   d.3,2,-3,+½

29. What are the n and l values for 4s and 5f orbitals?
   a.   4s: 4=4 l=0   b. 4s: n=5 l=0
      5f: n=4 l=3   5f: n=4 l=3
   c.4s: n=4 l=0   d. None   05f: n=5 l=3

30. The azimuthal quantum number l=3, the maximum number of electrons in that orbital will be?
   a. 2   b. 14   c. 10   d. 6

Chapter 2

PERIODIC TABLE

2.1 INTRODUCTION

Elements such as gold, silver, copper, lead and mercury have been known since antiquity, the first scientific discovery of an element occurred in 1649 when Hennig Brand discovered phosphorous. A significant amount of knowledge concerning the properties of elements and their compounds was acquired by chemists in the next 200 years. By 1869, a total of 63 elements had been discovered. At present 114 elements are known. As the number of known elements grew, scientists began to recognize patterns in properties and began to develop classification schemes. Attempts have been made from time to time to classify elements on the basis of their physical and chemical properties, so that elements behaving in a similar manner can be studied collectively.

The periodic table was first introduced in 1869-70 by the German chemist Lothar Meyer and the Russian Dmitri Mendeleyev. Mendeleyev observed that when elements are arranged in the order of increasing atomic weights, element with similar properties appear at regular intervals. The periodic table provided a framework for the organisation of vast amount of information with respect to chemical behaviour of elements. In general periodic table helps in better and systematic understanding of chemistry of elements and their compounds.

2.2 GENESIS OF PERIODIC CLASSIFICATION

As more and more elements were discovered in the early 1800s, chemists started looking for some relations in their physical and chemical properties. The elements were placed in lists with no particular order till this time. Initial classification of elements considered atomic weights. But later atomic numbers of the elements gained more significance.

Later with great efforts of Johann Dobereiner, John Newlands, Lother Meyer and Mendeleev, the arrangement of elements into more systematic and easier format was achieved, for better understanding of their chemical behaviour, as in the present modern periodic table.

❖ *Dobereiner's Law of Triads (1829)*

Dobereiner, a German chemist, noticed a set of three similar elements called Triads, when arranged in the order of increasing atomic weights, the atomic weight of the middle element was found to be the average of the atomic weights of the other two elements. This is called Dobereiner's law of Triads. Examples include Li-Na-K; S-Se-Te; Cl-Br-I, etc. The major drawback of this law was that it could be applied to a limited number of elements.

Lithium - Li	Sodium- Na	Potassium- K	Average atomic weight
7	23	39	$\dfrac{7 + 39}{2} = 23$

Sulphur- S	Selenium- Se	Tellurium- Te	Average atomic weight
32	79	127.5	$\dfrac{32 + 127.5}{2} = 79.25$

Chlorine- S	Bromine- Br	Iodine- I	Average atomic weight
35.5	79.9	127	$\dfrac{35.5 + 127}{2} = 81.25$

❖ *Newlands' Law of Octaves (1864)*

An English chemist John Newland proposed a sequence of elements arranged in the increasing order of atomic weights and it is found that every eighth element has the properties similar to that of first element. (9th element resembled with 2nd element and so on). This was called Newland's law of Octaves because it was like a musical scale in which the first note is the same as the eighth. Thus the properties of lithium are similar to those of sodium, which is the eighth element for lithium.

Li 1	Be 2	B 3	C 4	N 5	O 6	F 7
Na 8	Mg	Al	Si	P	S	Cl
K	Ca					

This Newland's Octave law was also discarded because it cannot be applied to heavier elements. Also the discovery of noble gases meant that element was no longer similar to the first.

❖ *Lother Meyer's Curve(1869)*

Julius Lother Meyer was a German chemist calculated the atomic volume of all the sixty known elements. He observed from the plot of atomic volumes against atomic weights of the various elements, that similar elements occupy similar positions. The most strongly electropositive alkali metals (Li, Na, K Rb, Cs) occupy the peaks on the curve, the less electropositive metals (Be, Mg, ca, Sr, Ba) occupy descending positions and the most electronegative elements occupy ascending positions on the curve. Later in 1869 Lother Meyer proposed a significant point that the physical properties of the elements are a periodic function of their atomic weights.

❖ *Mendeleev's law*

D I Mendeleev, a Russian chemist in 1869 found that if elements are arranged in the increasing order of atomic weights, their chemical properties vary in regular pattern. Mendeleev modified his proposal after knowing Lother Meyer observations and presented it as Mendeleev's periodic law which can be stated as,

"The physical and chemical properties of elements are periodic functions of their respective atomic weights".

Mendeleev organized his material in terms of the families of the known elements which displayed similar properties. Mendeleev's arrangement of elements in the increasing order of atomic weights is called Mendeleev's periodic table which was published in 1905 is shown below:

PERIODIC SYSTEM OF THE ELEMENTS IN GROUPS AND SERIES

SERIES	GROUPS OF ELEMENTS								
	0	I	II	III	IV	V	VI	VII	VIII
1		Hydrogen H 1.008							
2	Helium He 4.0	Lithium Li 7.03	Beryllium Be 9.1	Boron B 11.0	Carbon C 12.0	Nitrogen N 14.04	Oxygen O 16.00	Fluorine F 19.0	
3	Neon Ne 19.9	Sodium Na 23.5	Magnesium Mg 24.3	Aluminium Al 27.0	Silicon Si 28.4	Phosphorus P 31.0	Sulphur S 32.06	Chlorine Cl 35.45	
4	Argon Ar 39.9	Potassium K 39.1	Calcium Ca 40.1	Scandium Sc 44.1	Titanium Ti 48.1	Vanadium V 51.4	Chromium Cr 52.1	Manganese Mn 55.0	Iron Fe 55.9 Cobalt Co 59 Nickel Ni 59 (Cu)
5		Copper Cu 63.6	Zinc Zn 65.4	Gallium Ga 70.0	Germanium Ge 72.3	Arsenic As 75	Selenium Se 79	Bromine Br 79.95	
6	Krypton Kr 81.8	Rubidium Rb 85.4	Strontium Sr 87.6	Yttrium Y 89.0	Zirconium Zr 90.6	Niobium Nb 94.0	Molybdenum Mo 96.0		Ruthenium Ru 101.7 Rhodium Rh 103.0 Palladium Pd 106.5 (Ag)
7		Silver Ag 107.9	Cadmium Cd 112.4	Indium In 114.0	Tin Sn 119.0	Antimony Sb 120.0	Tellurium Te 127.6	Iodine I 126.9	
8	Xenon Xe 128	Caesium Cs 132.9	Barium Ba 137.4	Lanthanum La 139	Cerium Ce 140				
9									
10				Ytterbium Yb 173		Tantalum Ta 183	Tungsten W 184		Osmium Os 191 Iridium Ir 193 Platinum Pt 194.9 (Au)
11		Gold Au 197.2	Mercury Hg 200.0	Thallium Tl 204.1	Lead Pb 206.9	Bismuth Bi 208			
12			Radium Ra 224		Thorium Th 232		Uranium U 239		
	R	R_2O	RO	R_2O_3	HIGHER SALINE OXIDES				
					RO_2	R_2O_5	RO_3	R_2O_7	RO_4
					HIGHER GASEOUS HYDROGEN COMPOUNDS				
					RH_4	RH_3	RH_2	RH	

Mendeleev's Periodic Table published earlier

Mendeleev could propose a periodic table where the then known 63 elements were arranged in the increasing order of the atomic weights. It consists of 8 vertical columns called groups and 7 horizontal rows called periods. Zero group was added to this periodic table after the discovery of noble gases.

- **Vertical columns**: Mendeleev's periodic table consists of nine vertical columns called groups. These groups are named as I, II, III, IV, V, VI, VII, VIII and 0 groups. Every group, from I to VII is divided into two sub groups A and B. Group VIII and 0 have no sub groups. Group VIII contains nine elements in sets of three and 0 group contain noble gases.

- **Horizontal rows**: horizontal rows are called periods. The table contains seven periods. The first period contains only 2 elements, the second and third period contain 8 elements each. The fourth, fifth and sixth periods contain 18 elements. The seventh period is incomplete.

Mendeleev relied on the similarities in the empirical formulas and properties of compounds formed by the elements. From the gaps present in his table, Mendeleev predicted the existence and properties of unknown elements which he called eka-aluminum and eka-silicon. Mendeleev in order to give provisional names to his predicted elements, used the prefixes eka-, dvi-, and tri-, from the Sanskrit names of digits 1, 2, and 3, depending upon whether the predicted element was one, two, or three places down from the known element of the same group in his table. For example, germanium was called eka-silicon until its discovery in 1886, and rhenium was called dvi-manganese before its discovery in 1926.The elements gallium and germanium were found later to fit his predictions quite well. Some of the properties predicted by Mendeleev for eka-aluminum and eka-silicon and those found experimentally are given below:

Property	Eka-aluminium (Predicted)	Gallium (Found)	Eka-Silicon (Predicted)	Germanium (Found)
Atomic weight	68	70	72	72.6
Density (g/cm^3)	5.9	5.94	5.5	5.36
Melting point (K)	Low	302.9	high	1231
Formula of oxide	E_2O_3	Ga_2O_3	EO_2	GeO_2

Merits and demerits- Mendeleev's periodic table

❖ Merits of Mendeleev's periodic classification

1. Classification of all elements: Mendeleev's was the first classification which successfully included all the elements.
2. Prediction of new elements: Mendeleev's periodic table had some blank spaces in it. These vacant spaces were for elements that were yet to be discovered. For instance, scandium, gallium and germanium, discovered later, have properties similar to Eka–boron,Eka–aluminium and Eka–silicon, respectively.

❖   Demerits of Mendeleev's periodic classification

1. *Position of hydrogen*: Hydrogen resembles alkali metals as well as halogens. Therefore, it could neither be placed with alkali metals nor with halogens (group VII).
2. *Position of isotopes*: Different isotopes of same elements have different atomic masses, therefore, each one of them should be given a different position in the periodic table. On the other hand, because they are chemically similar, they had to be given same position.
3. *Anomalous pairs of elements*: At certain places, an element of higher atomic mass has been placed before an element of lower atomic mass. For example, Argon (39.91) is placed before potassium(39.1).

2.3 MODERN PERIODIC TABLE

In the year 1913, Henry Moseley discovered a new property of elements, called Atomic number, which provided a better basis for the periodic arrangement of the elements. Mendeleev periodic law was modified and called as Modern periodic law.

The last major changes to the periodic table resulted from Glenn Seaborg's work in the middle of the 20th Century. Starting with his discovery of plutonium in 1940, he discovered all the transuranic elements from 94 to 102. He reconfigured the periodic table by placing the actinide series below the lanthanide series. In 1951, Seaborg was awarded the Nobel Prize in chemistry for his work. Element 106 has been named seaborgium (Sg) in his honor.

The Modern periodic law states that **"The physical and chemical properties of elements are periodic function of their atomic numbers"**.

The periodic table: A table in which elements are arranged in the order of increasing atomic numbers in the manner that the elements with similar properties fall in the same vertical column is known as the periodic table.

The elements, which fall in the same column and resemble one another in their properties are said to belong to the same group or family of elements.

Importance of the periodic table
- The objective of the periodic table is to organize and systematize the chemistry of the elements.
- It helps us to understand the characteristic behaviour of each element and resemblance of group elements.
- It helps us to understand the cause of periodicity of properties and the reason why similar properties recur at certain regular intervals, i.e., after 2,8,18 and 32 elements. These numbers namely 2,8,18 and 32 sometimes referred to as magic numbers.

The long form of the periodic table

It is an arrangement in which the elements are arranged in the increasing order of atomic number such that the elements with similar properties fall in vertical columns called **groups:** Horizontal rows are called **periods.** Periodic table consists of 18 groups and 7 periods.

Instead of listing the 103 elements as one long list, the periodic table arranges them into several horizontal rows or periods, in such a way that each row begins with an alkali metal and ends with a noble gas. The sequence in which the various energy levels are filled determines the number of elements in each period and the periodic table can be divided into four main regions according to whether the s, p, d or f-levels are being filled.

Periods			No. of elements in a period
1	1s	→	2
2	2s 2p	→	8
3	3s 3p	→	8
4	4s 3d 4p	→	18
5	5s 4d 5p	→	18
6	6s 4f 5d 6p	→	32
7	7s 5f 6d 7p	→	30 incomplete

Characteristics of groups

- All the group elements have same valence electronic configuration and thus exhibits similar chemical properties.
- The elements in a group are separated by definite gaps of atomic numbers.(2,8,8,18,18,32- Magic numbers)
- In general physical properties of group elements (melting point, boiling point etc) follow a systematic pattern down the group.
- The atomic sizes of the elements in a group always decrease down the group.
- s-block elements belong to group 1 and 2, p-block elements belong to groups 13 to 18, d-block elements belong to groups 3 to 12 and f-block elements are considered to be in group 3.

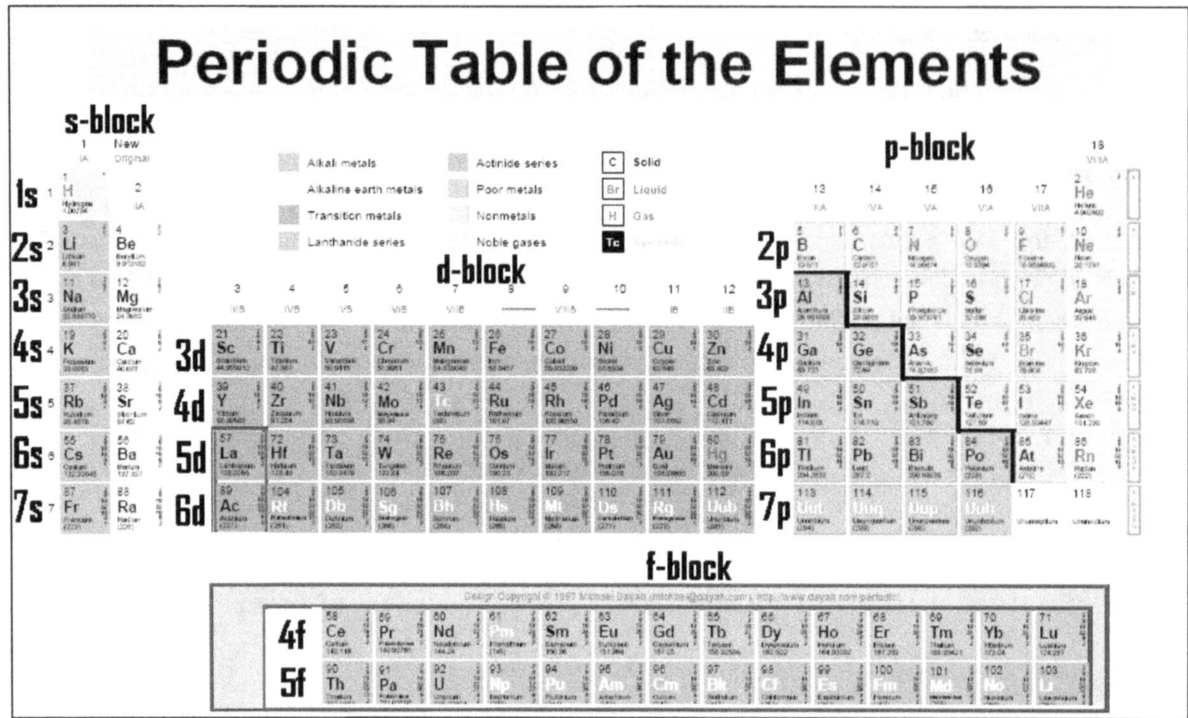

Characteristics of periods

- In all the elements of period, electrons are successfully filled to the same valence shell and thus all the period elements have different electronic configuration.

- In general elements exhibit different chemical properties along a period.

- The metallic character of elements decreases across a period, while their non metallic character increases.

Cause of periodicity

The cause of periodicity in properties appears to lie in the recurrence of similar outer electronic configurations at certain regular intervals. For example, the general electronic configuration of first group alkali metals is ns^1. The recurrence of the same electronic configuration for alkali metals, starts from Na[$3s^1$ after Li [$2s^1$]] then continues for K[$4s^1$], Rb [$5s^1$] Cs[$6s^1$] Fr [$7s^1$] .

The similar outer electronic configurations recur at definite intervals of atomic numbers 2, 8, 8, 18, 18, 32 and 32. Hence periodicity in properties appears.

2.4 NOMENCLATURE OF ELEMENTS

Nomenclature of Elements of Atomic Numbers greater than 100

The naming of element, according to IUPAC, was based on a property of the element, a mineral from which it was isolated, its place or area of discovery, a mythological character or concept, an astronomical object, or name of an eminent scientist to honor his work. The nomenclature for elements till atomic number 103 was a smooth phenomenon. But after 103, naming elements became a tough task when different persons laid claims on the discovery of some of these elements.

Elements of atomic numbers greater than 103 are often referred to in the scientific literature but receive names only after they have been discovered. Names are needed for these elements even before their existence has been established and therefore a commission on nomenclature of inorganic chemistry-CNIC (Appointed by IUPAC) has approved a systematic nomenclature and series of three-letter symbols for the atoms of such elements.

Nomenclature of Elements of Atomic Numbers greater than 100

Based on the suggestions by CNIC, the following nomenclature of elements of atomic numbers greater than 100 was derived:

1. The name is derived directly from the atomic number of the element using the following numerical roots

0 = nil	3 = tri	6 = hex
1 = un	4 = quad	7 = sept 9 = enn
2 = bi	5 = pent	8 = oct

2. The roots are put together in the order of the digits which make up the atomic number and terminated by "ium" to spell out the name. The final "n" of "enn" is elided when it occurs before "nil", and the final "i" of "bi" and of "tn" when it occurs before "ium".

3. The symbol of the element is composed of the initial letters of the numerical roots which make up the name.

4. The root "un" is pronounced with a long "u", to rhyme with "moon". In the element names each root is to be pronounced separately.

For example Atomic number 118 has three numerical roots: un, un and oct. The final root must be "ium", so the name becomes un+un+oct+ium = **ununoctium**. Hence nomenclatures of elements of atomic numbers greater than 100 are as given below:

Atomic number	Name		Symbol
101	Mendelevium (Unnilunium)		Md*
102	Nobelium (Unnilbium)		No*
103	Lawrencium (Unniltrium)		Lr*
104	Unnilquadium	Ruther fordium-Rf	Unq
105	Unnilpentium	Dubniun-Db	Unp
106	Unnilhexium	Seaborgium-Sg	Unh
107	Unnilseptium	Bohrium-Bh	Uns
108	Unniloctium	Hassnium-Hs	Uno
109	Unnilennium	Meitnerium-Mt	Une
110	Ununnilium	Darmstadium-Ds	Uun
111	Unununium	Roentgenium-Rg	Uuu
112	Ununbium		Uub

113	Ununtrium	Uut
114	Ununquadium	Uuq
115	Ununpentium	Uup
116	Ununhexium	Uuh
117	Ununseptium	Uus
118	Ununoctium	Uuo
119	Ununennium	Uue
120	Unbinilium	Ubn
121	Unbiunium	Ubu
130	Untrinilium	Utn
140	Unquadnilium	Uqn
150	Unpentnilium	Upn
160	Unhexnilium	Uhn
170	Unseptnilium	Usn
180	Unoctnilium	Uon
190	Unennilium	Uen
200	Binilnilium	Bnn

2.5 *s, p, d* AND *f* BLOCK ELEMENTS

The aufbau (build up) principle and the electronic configuration of atoms provide a theoretical foundation for the periodic classification. The elements in a vertical column of the Periodic Table constitute a group or family and exhibit similar chemical behaviour. This similarity arises because these elements have the same number and same distribution of electrons in their outermost orbitals. The periodic table is divided into four blocks namely **s, p, d and f blocks** depending on the type of atomic orbitals that are being filled with electrons.

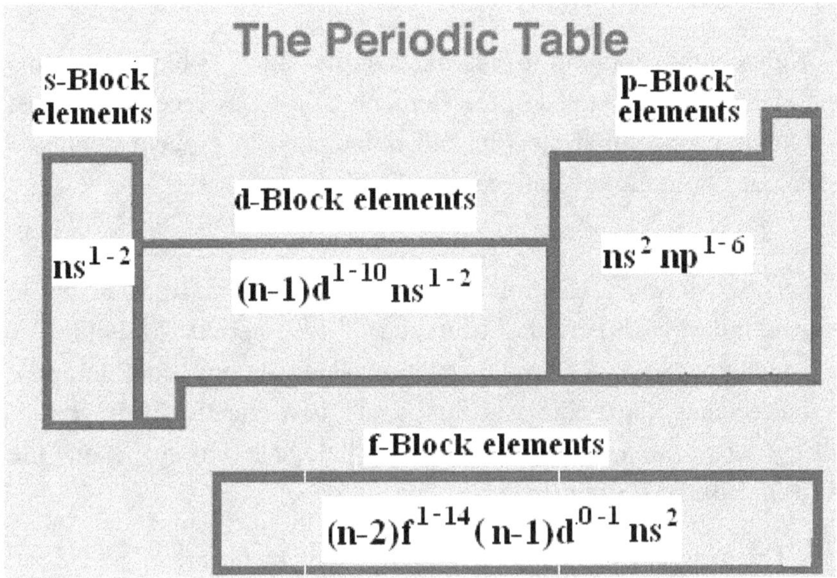

The element in any block is linked with the sub-shell or orbital in which the last electron in its atom is filled. If the valence electron is present in s-orbital then it belongs to s-block. Same is the case with the other elements.

The S-block elements:

The general electronic configuration of elements belonging to this block is

ns^{1-2}. n is principal quantum number. S blocks include 1 and 2 groups.

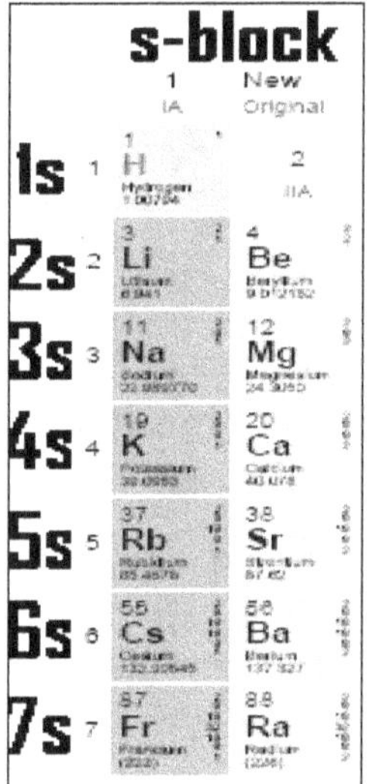

Group 1 elements (Alkali metals): Li, Na, K, Rb, Cs & Fr.

Group 2 elements (Alkaline earth metals): Be, Mg, Ca, Sr, Ba and Ra.

The atoms of elements of the two groups 1 and 2 receive the last electron

in the S-orbital of their outermost energy level.

The general characteristics of s- block elements are:

1. They are soft metals.
2. They are highly reactive sand form ionic compounds.
3. They have low ionization enthalpies and are highly electropositive.
4. They have +1 and +2 oxidation states.
5. They act as good reducing agents as they lose electron readily.

Note*   *Hydrogen which is placed above Group 1 alkali metals is not a member of that group because it is neither metallic nor a solid.*

The P-block elements:

The general electronic configuration of elements belonging to this block is $ns^2\ np^{1-6}$. P-block elements include 13, 14, 15, 16, 17 and 18 group elements. The Atoms of the elements of p-block receive their last electron in their p-orbitals. Helium contains no p orbital electron but is included in group 18. This is because helium completes k- Shell electronic configuration and exhibit chemical inertness.

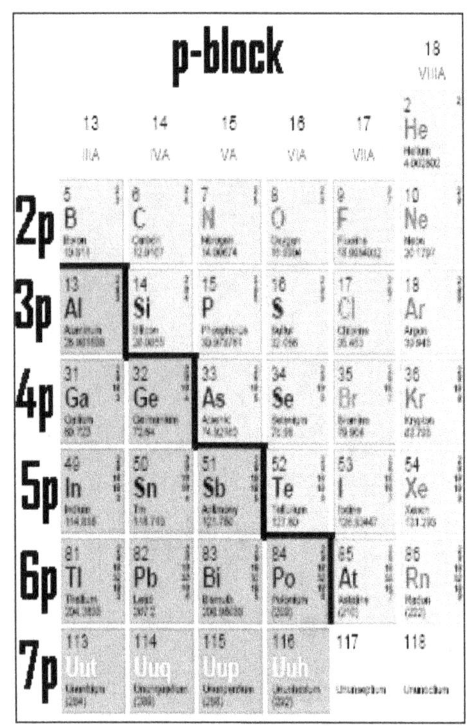

The outermost electronic configuration varies from ns^2np^1 to ns^2np^6 in each period. At the end of each period is a noble gas element with a closed valence shell ns^2np^6 configuration. Preceding the noble gas family are two chemically important groups of non-metals. They are the halogens (Group 17) and the chalcogens (Group 16).

The general Characteristics of p-block elements are:
1. p-block elements are representative elements which include metals, metalloids and non metals.
2. Compounds of these elements are mostly covalent in nature.
3. They exhibit variable oxidation states
4. They are highly electronegative elements.
5. Most of them form acidic oxides.
6. The nonmetallic character increases left to right across the periods and metallic character increases down the groups.
7. The reactivity of the elements decreases down the group.

The d-block (Transition) elements

The general electronic configuration of elements of d block is $(n-1)d^{1-10}\ ns^{1-2}$. The d-orbitals in the penultimate [inner to the outermost] shell, however, are still empty. The last electron in these elements therefore, enters into the d-orbital, which thereby, get progressively filled as we move across a period. There are four transition series in d-block. Thus the elements from:

<1>   Sc to Zn   - 3d series

<2>   Y to Cd   - 4d series

<3>   La to Hg   - 5d series,

<4>   Ac to Uub - 6d series are called d-block elements.

The d-block elements are known as transition elements because the properties of d- block elements are transitional (Intermediate) between s-block elements and p-block elements.

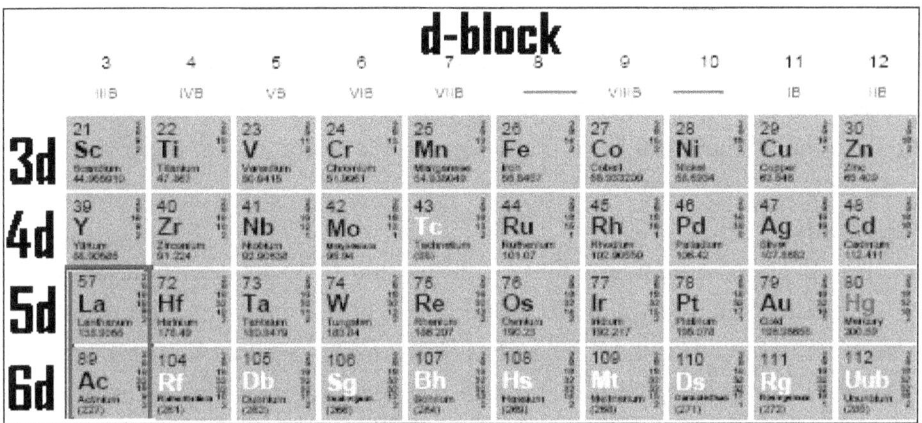

The general characteristics of d-block elements are

1. They are all metals with high melting and boiling points.
2. They form coordinate complexes which give coloured solutions.
3. They exhibit variable oxidation states
4. Most of the compounds of elements exhibit paramagnetic nature.
5. They possess good catalytic properties.

Elements of group 12 (Zn, Cd and Hg) donot show transition properties, because d- orbital is fully filled.

The f-block elements

General electronic configuration is $(n-2) f^{1-14} (n-1) d^{0-1} ns^2$. The elements in which the last electron enters into the f-orbitals of their atoms are called f-block elements. f-blocks include 14 lanthanides and 14 Actinides.

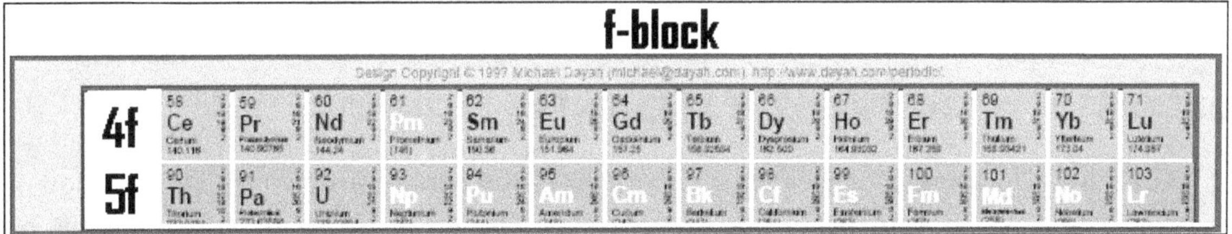

In lanthanides, 4-f orbitals and in actinides, 5-f orbitals are being progressively filled. The two rows of elements at the bottom of the Periodic Table, called the Lanthanoids, Ce(Z = 58) – Lu(Z = 71) and Actinoids, Th(Z = 90) – Lr (Z = 103). In this f-block, atoms of the elements receive electrons to their pre-penultimate shell (innermost), therefore f-block elements are also called inner transition elements. **The elements after uranium are called** Transuranium elements.

Position of hydrogen

Hydrogen is more difficult to place in a group. It could be included in group I because it has one s electron in its outer shell, is univalent and commonly forms univalent positive ions. However, hydrogen is not a metal and is a gas while Li, Na, K, Rb, and Cs are metals and solids. Similarly, hydrogen could be included in Group VII because it is one electron short of a complete shell or in Group IV because its outer shell is half full. Hydrogen doesn't resemble the alkali metals, the halogens or group IV very closely. Hydrogen atoms are extremely small and have many unique properties. Thus there is a case for placing hydrogen in a group on its own.

Metals, Non-metals and Metalloids:

In addition to displaying the classification of elements into s-, p-, d-, and f-blocks, periodic table shows another broad classification of elements based on their properties. The elements can be divided into Metals and Non-Metals. Metals comprise more than 78% of all known elements and appear on the left side of the Periodic Table. Metals are usually solids at room temperature [mercury is an exception; gallium and caesium also have very low melting points (303K and 302K, respectively)]. Metals usually have high melting and boiling points. They are good conductors of heat and electricity. They are malleable (can be flattened into thin sheets by hammering) and ductile (can be drawn into wires).

In contrast, non-metals are located at the top right hand side of the Periodic Table. In fact, in a horizontal row, the property of elements changes from metallic on the left to non-metallic on the right. Non-metals are usually solids or gases at room temperature with low melting and boiling points (boron and carbon are exceptions). They are poor conductors of heat and electricity. Most nonmetallic solids are brittle and are neither malleable nor ductile. The elements become more metallic as we go down a group; the nonmetallic character increases as one goes from left to right across the Periodic Table. The change from metallic to non-metallic character is not abrupt as shown by the thick zig-zag line. The elements (e.g., silicon, germanium, arsenic, antimony and tellurium) bordering this line and running diagonally across the Periodic Table show properties that are characteristic of both metals and nonmetals. These elements are called Semi-metals or Metalloids.

2.6 PERIODIC TRENDS IN PROPERTIES OF ELEMENTS

The physical and chemical properties of elements, either across the period or down the group, show a definite periodic variation which is probably due to systematic change in electronic configurations of elements which are repeated after intervals of 2, 8, 8, 18, 18 and 32. This is responsible for periodicity in certain properties of elements such as atomic radius, ionic radius, ionization enthalpy, electron gain enthalpy and electronegativity.

Trends in physical properties:

❖ **a) Atomic radius**

In general, the distance between the centre of the nucleus and the outermost shell of electrons is called radius of atom. However, there is no certainty with regard to the exact position occupied by electrons at any time. It may be very close to the nucleus at one time or may be for a way from the nucleus at next moment.

There are three important concepts of atomic radius.

i) **Vander Waals Radius: One half of the distance between the nuclei of two adjacent atoms belonging to two neighbouring molecules of an element in the solid state is usually known as vander waals radius (Fig.9).**

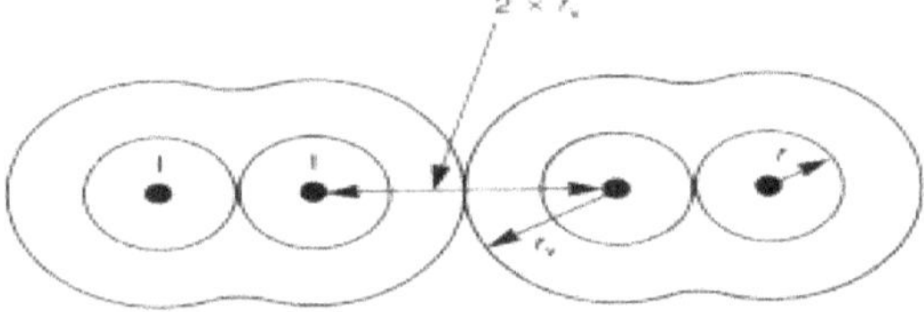

Fig. 9. Atomic radius

ii) **Covalent radius**: **One half of the distance between the nuclei of two covalently bonded atoms of the same element in a molecule is taken as the covalent radius of the atom of that element.**

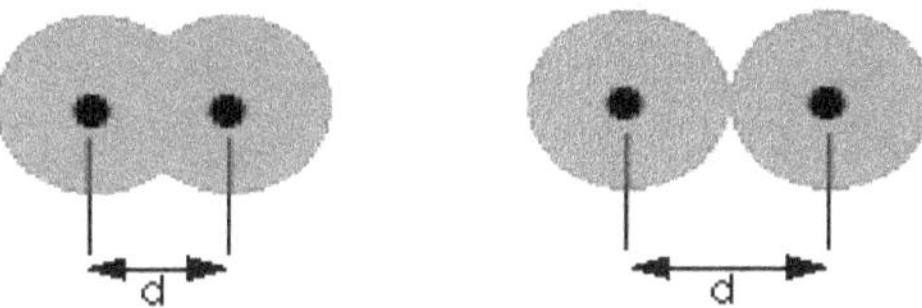

Covalent radius of chlorine atom is = 99 pm

iii) Metallic radius
Metallic radius is taken as half of the internuclear distance separating the metal cores in the metallic crystal. For example the distance between two adjascent copper atoms in solid copper is 256 pm; hence the metallic radius of copper is assigned a value of 128 pm.

❖ **Variation of atomic radii in a period**
Atomic number increases on moving from left to right across a period. Nuclear charge of atoms also increases across a period. As the atomic number increases, the effective nuclear charge increases. The electrons which are added to the same orbital across a period, experience a least shielding by the electrons already in the orbital and experience more nuclear charge, Hence these electrons are getting attracted more and more towards the nucleus. Therefore size of an atom reduces on moving across a period. The atomic radius decreases on moving left to right across a period (Table 1).

Table 1   Atomic Radii/pm Across the Periods

ATOM (PERIOD II)	Li	Be	B	C	N	O	F
ATOMIC RADIUS	152	111	88	77	74	66	64
ATOM (PERIOD III)	Na	Mg	Al	Si	P	S	Cl
ATOMIC RADIUS	186	160	143	117	110	104	99

Trends in atomic radius in Periods 2 and 3

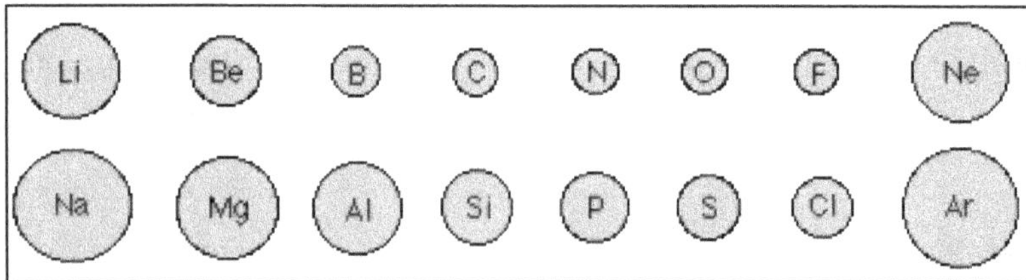

Trends in atomic radius across periods

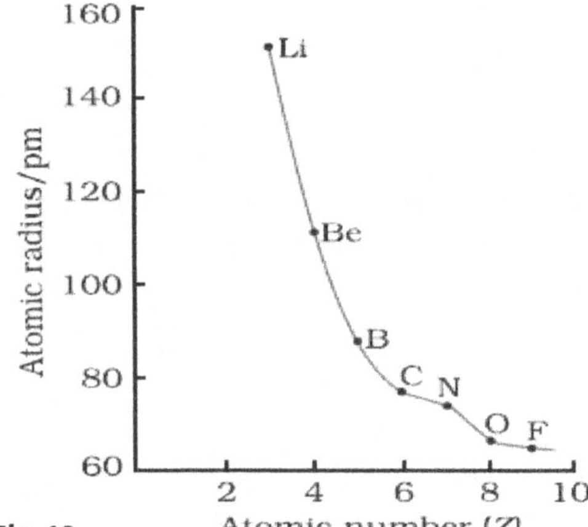

Fig. 10.
Variation of atomic radius with atomic number across the second period

The atomic size generally decreases across a period as illustrated in Fig. 10, for the elements of the second period. It is because within the period the outer electrons are in the same valence shell and the effective nuclear charge increases as the atomic number increases resulting in the increased attraction of electrons to the nucleus.

You have to ignore the noble gas at the end of each period. Because neon and argon don't form bonds, you can only measure their van der Waals radius - a case where the atom is pretty well "unsquashed". All the other atoms are being measured where their atomic radius is being lessened by strong attractions.

Variation of atomic radii in a group:
Atomic radius increases on moving down a group. In moving down a group, the number of shells increases and therefore, the size of the atom increases. This effect no doubt, is partly annulled by the drawing in of the electron shell on account of the increasing nuclear charge. But the effect of adding a new shell is so large that it overcomes the contractive effect of the increased nuclear charge. Hence the radius of the atom increases in moving from top to bottom in a group (Table 2).

Table 2 **Atomic Radii/pm Down a Family**

ATOM(GROUP I)	ATOMIC RADIUS	ATOM (GROUP 17)	ATOMIC RADIUS
Li	152	F	64
Na	186	Cl	99
K	231	Br	114
Rb	244	I	133
Cs	262	At	140

Trends in atomic radius down a group

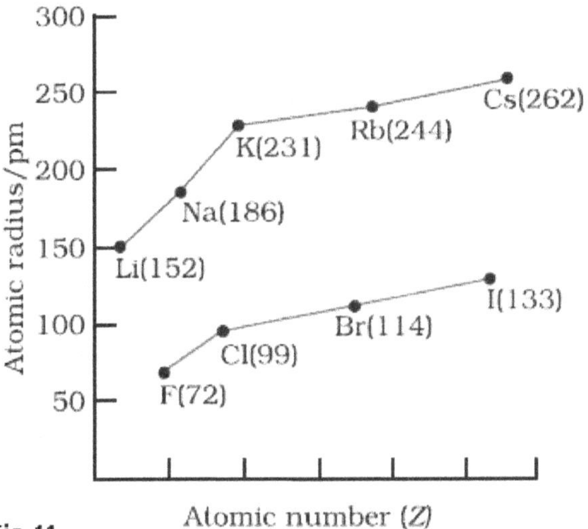

Fig. 11.
Variation of atomic radius with atomic number for alkali metals and halogens

Within a family or vertical column of the periodic table, the atomic radius increases regularly with atomic number as illustrated in Fig. 11. For alkali metals and halogens, descending the groups, the principal quantum number (n) increases and the valence electrons are farther from the nucleus. This happens because the inner energy levels are filled with electrons, which serve to shield the outer electrons from the pull of the nucleus. Consequently the size of the atom increases as reflected in the atomic radii. It is fairly obvious that the atoms get bigger as you go down groups. The reason is equally obvious - you are adding extra levels of electrons.

❖ **b) Ionic radius.** Ionic radius may be defined as the distance from the nucleus of an ion up to which it has influence on its electron cloud. The internuclear distance between the nuclei of Na^+ and Cl^- ions is determined from x-ray technique. This distance is taken as the sum of the radii of Na^+ and Cl^- ions. The contribution of Na^+ and Cl^- ions to the internuclear distance is then evaluated. In this way, the radii of Na^+ and Cl^- found out.

Radii of Cations and Anions

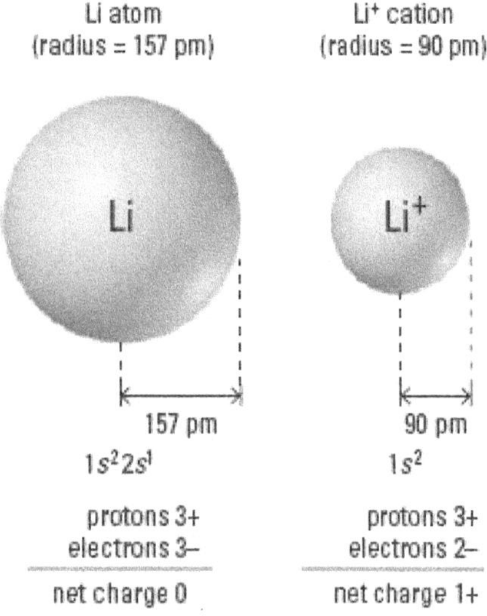

Fig. 12. Radius of lithium ion

A positive ion is obtained by the loss of one or more electrons from the outer shell or valence shell of an atom (Fig. 12). In the formation of positive ions, the outer shell of electrons is generally removed completely. The cation, therefore is much smaller than the corresponding atom. At the same time, with the elimination of one or more orbital electrons. The effective nuclear charge increases. The electrons are pulled in more towards the nucleus than before. This effect also tends to decrease the radius of the cation.

- **The radius of the cation is always smaller than that of the atom.**

Atomic radius pm	Ionic radius pm
Na = 186	$Na^+ = 9$
Al = 143	$Al^{3+} = 50$
Mg = 160	$Mg^{2+} = 80$

It is evident that radius of a cation is invariably smaller than that of the corresponding atom. Addition of one or more electrons into an outermost shell of a neutral atom leads to the formation of an anion. The effective nuclear charge of anion of atom is reduced and therefore, the electrons acquire greater freedom of getting away from the nucleus. As a result of this, the radius of a negative ion is invariably larger than that of the corresponding atom

Fig. 13. Radius of fluoride ion

- **The radius of the anion is always greater than that of the atom.**

Atomic radius pm	Ionic radius pm
Cl - 99	Cl⁻ - 198
Br - 114	Br⁻ - 195

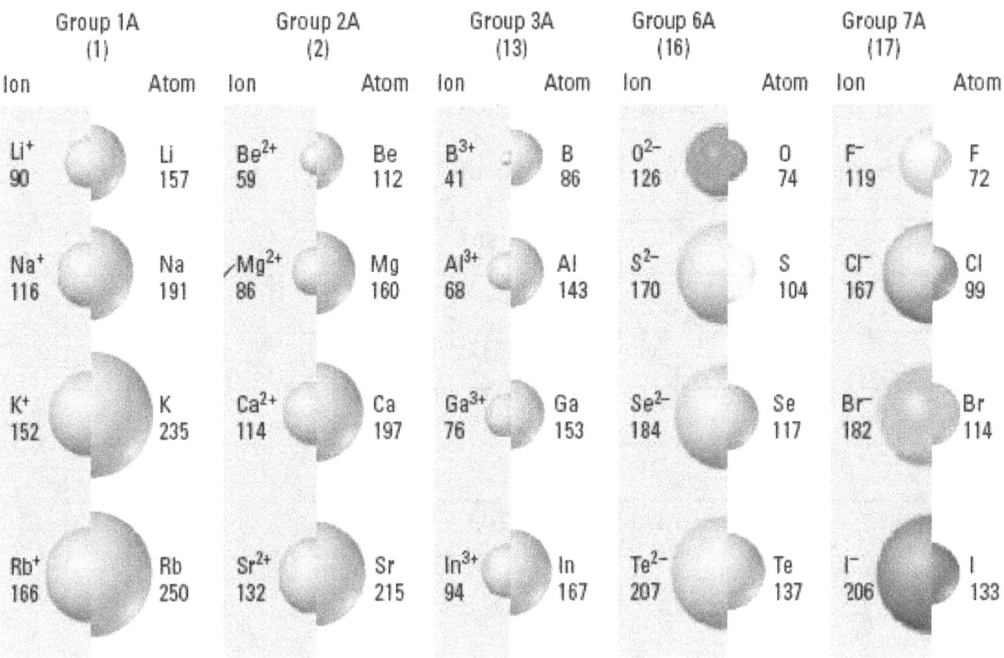

Fig. 14. Atomic and atomic radii

Isoelectronic ions: Isoelectronic ions are those ions which have the same number of electrons but differ in the charge on their nuclei.

Table 3. Radii of isoelectronic species

Ions	N^{3-}	O^{2-}	F^-	Na^+	Mg^{2+}	Al^{3+}
No. of electrons	10	10	10	10	10	10
Charge on nucleus	7	8	9	11	12	13
Radius pm	171	140	136	95	80	50

c) Ionisation enthalpy $(\Delta_i H)$

Ionisation enthalpy is the measure of the tendency of an atom of an element to lose electron. The ionisation enthalpy of an element is the minimum amount of energy required to remove an electron from an isolated gaseous atom of that element resulting in the formation of a positive ion.

Ionisation enthalpy ($\Delta_i H$) is the measure of the tendency of an atom to lose an electron and to become a cation, as represented below:

$$X(g) \xrightarrow{I.E1} X^+(g) + e \qquad X^+(g) \xrightarrow{IE_2} X^{2+}(g) + e$$

$$X^{2+}(g) \xrightarrow{IE_3} X^{3+}(g) + e$$

The energy required to remove the first electron from an atom is called the first ionisation enthalpy IE_1, the energies required to remove the second, third electrons etc are called second IE [IE_2] third I.E[IE_3] etc. Energy is always required to remove electrons from an atom and hence ionization enthalpies are always positive. The second ionization enthalpy will be higher than the first ionization enthalpy because it is more difficult to remove an electron from a positively charged ion than from a neutral atom. In the same way the third ionization enthalpy will be higher than the second and so on. The term "ionization enthalpy", if not qualified, is taken as the first ionization enthalpy.

Table 4. Ionisation enthalpies for group I and II elements [KJ mol^{-1}]

	1st	2nd		1st	2nd	3rd
Li	520	7296	Be	899	1757	14847
Na	496	4563	Mg	737	1450	7731
K	419	3069	Ca	590	1145	4910
Rb	403	2650	Sr	549	1064	4207
Cs	374	2420	Ba	503	965	
Fr	------	--------	Ra	509	979	

The factors that influence the ionisation enthalpy are:
1. The size of the atom
2. The charge on the nucleus
3. How effectively the inner electron shells screen the nuclear charge?- shielding effect
4. The type of electron involved (s.p.d or f)

In a small atom the electrons are tightly held, whilst in a larger atom, the electrons are loosly held. Thus the I.E decreases as the size of the atoms increases.

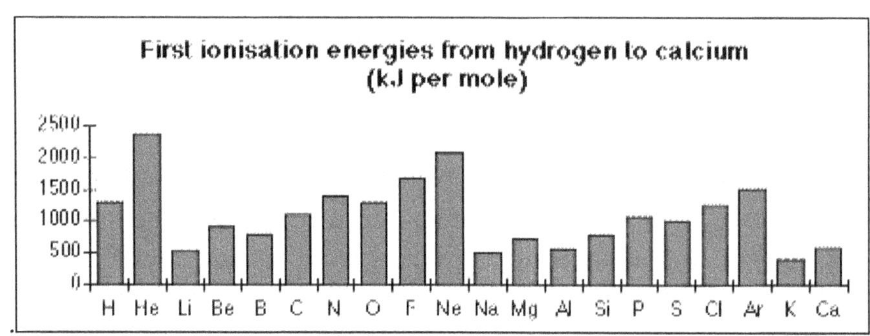

First ionisation energy shows *periodicity*. That means that it varies in a repetitive way as you move through the Periodic Table. For example, look at the pattern from Li to Ne, and then compare it with the identical pattern from Na to Ar.

These variations in first ionisation enthalpy can all be explained in terms of the structures of the atoms involved (Fig.15). The ionisation enthalpy also depends on the type of electron, which is being removed. s,p,d and f electrons have orbitals with different shapes. An s electron penetrates nearer to the nucleus and is therefore more tightly held than a p electron. For similar reasons a p electron is more tightly held than a d electron and d electron is more tightly held than f electron. Other factors being equal the ionisation enthalpies are in the order s>p>d>f. Thus the increase in I.E is not quite smooth on moving from left to right in the periodic table.

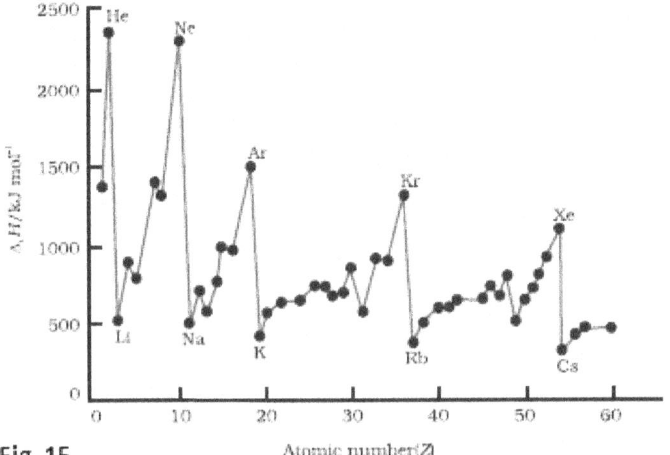

Fig. 15.

Variation of first ionization enthalpies $(\Delta_i H)$ with atomic number for elements with Z = 1 to 60

In general, the IE decreases on descending a group and increases on crossing a period. Removal of successive electrons becomes more difficult and first IE < second IE < third IE. The first ionization enthalpies of elements having atomic numbers up to 60 are plotted in the figure. The periodicity of the graph is quite striking. Maxima is observed at the noble gases which have closed electron shells and very stable electron configurations. On the other hand, minima occur at the alkali metals and their low ionization enthalpies can be correlated with their high reactivity. In general first ionization enthalpy increases as we go across a period and decreases as we descend in a group.

To understand these trends, we have to consider two factors : (i) the attraction of electrons towards the nucleus, and (ii) the repulsion of electrons from each other. The effective nuclear charge experienced by a valence electron in an atom will be less than the actual charge on the nucleus because of "shielding" or "screening" of the valence electron from the nucleus by the intervening core electrons. For example, the 2s electron in lithium is shielded from the nucleus by the inner core of 1s electrons. As a result, the valence electron experiences a net positive charge which is less than the actual charge of +3.

In general, shielding is effective when the orbitals in the inner shells are completely filled. This situation occurs in the case of alkali metals which have a lone ns-outermost electron preceded by a noble gas electronic configuration. On moving from lithium to fluorine across the second period, successive electrons are added to orbitals in the same principal quantum level and the shielding of the nuclear charge by the inner core of electrons does not increase very much to compensate for the increased attraction of the electron to the nucleus. Thus, **across a period, increasing nuclear charge outweighs the shielding.** Consequently, the outermost electrons are held more and more tightly and the **ionization enthalpy increases across a period (Fig.16).** As we go down a group, the outermost electron being increasingly farther from the nucleus, there is an increased shielding of the nuclear charge by the electrons in the inner levels. In this case, **increase in shielding outweighs the increasing nuclear charge and the removal of the outermost electron requires less energy down a group.**

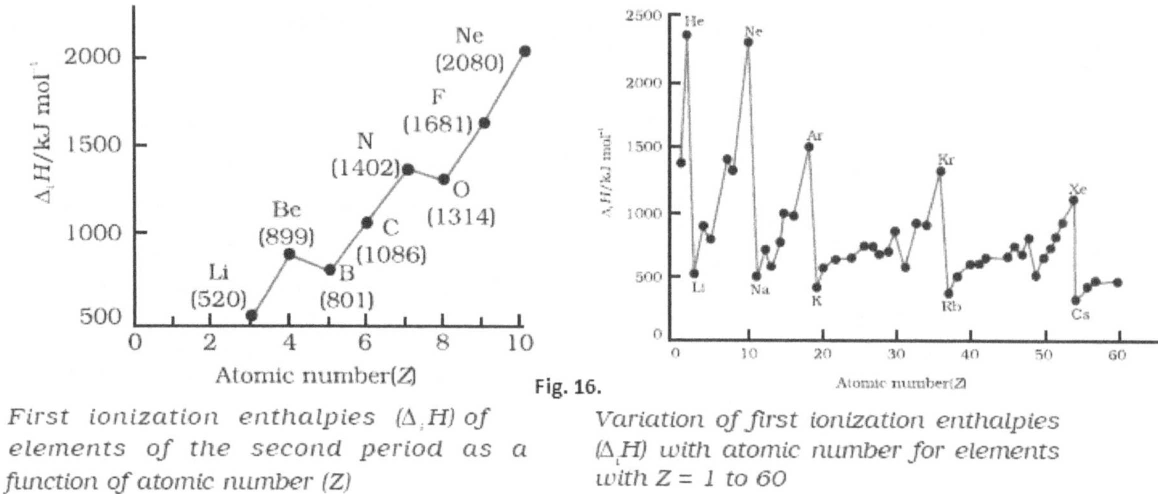

Fig. 16.

First ionization enthalpies ($\Delta_i H$) of elements of the second period as a function of atomic number (Z)

Variation of first ionization enthalpies ($\Delta_i H$) with atomic number for elements with Z = 1 to 60

d) Electron gain enthaply ($\Delta_{eg} H$)

Electron gain enthalpy is the measure of the tendency of an atom of an element to accept electron. The energy released when an extra electron is added to a neutral gaseous atom to convert it in to its anion is termed the electron gain enthalpy.

The electron affinity gives the measure of the tendency of an atom to change in to anion.

$$X(g) + e \longrightarrow X^-(g)$$

Usually only one electron addition gives a uni negative ion. Since energy is evolved, these terms have a negative sign. Electron affinities depend on the size and effective nuclear charge. Negative electron affinity values indicate that energy is given out when the atom accepts an electron. The above values show that the halogens evolve a large amount of energy on forming negative halide ions.Energy is evolved when one electron is added to an O and s atom, forming the species O^- and S^-, but a substantial amount of energy is absorbed when two electrons are added to form O^{2-} and S^{2-} ions. Thus the electron affinities to $O \rightarrow O^{2-}$ and $S \rightarrow S^{2-}$ have a positive side. group 17 elements (the halogens) have very high negative electron gain enthalpies because they can attain stable noble gas electronic configurations by picking up an electron. On the other hand, noble gases have large positive electron gain enthalpies because the electron has to enter the next higher principal quantum level leading to a very unstable electronic configuration. In general, the electron affinity decreases in going from top to bottom in a group and increases in going from left to right across a period (Table 5).

Table 5 Electron Gain Enthalpies / (kJ mol^{-1}) of Some Main Group Elements

GROUP I	$\Delta_{EG}H$	GROUP 16	$\Delta_{EG}H$	GROUP 17	$\Delta_{EG}H$	GROUP 0	$\Delta_{EG}H$
H	-73					He	+48
Li	-60	O	-141	F	-328	Ne	+116
Na	-53	S	-200	Cl	-349	Ar	+ 96
K	-48	Se	-195	Br	-325	Kr	+ 96
Rb	-47	Te	-190	I	-295	Xe	+ 77
Cs	-46	Po	-174	At	-270	Rn	+ 68

Variation of electron affinity along the period

The variation in electron gain enthalpies of elements is less systematic than for ionization enthalpies. As a general rule, electron gain enthalpy becomes more negative with increase in the atomic number across a period. The effective nuclear charge increases from left to right across a period and consequently it will be easier to add an electron to a smaller atom since the added electron on an average would be closer to the positively charged nucleus. However, electron gain enthalpy of O or F is less negative than that of the succeeding element. This is because when an electron is added to O or F, the added electron goes to the smaller n = 2 quantum level and suffers significant repulsion from the other electrons present in this level. For the n = 3 quantum level (S or Cl), the added electron occupies a larger region of space and the electron-electron repulsion is much less.

Variation of electron affinity down the group

Electron affinity decreases down the group due to increase in the atomic size and decrease in the nuclear attraction for electrons. On moving down a group, electron gain enthalpy become less negative because, the size of the atom increases and the added electron would be farther away from the nucleus.

e) Electro negativity

Electronegativity is a measure of the tendency of an atom to attract a shared pair of electrons to itself

The Pauling scale is the most commonly used. Fluorine (the most electronegative element) is assigned a value of 4.0, and values range down to caesium and francium which are the least electronegative at 0.7. The electronegativity of any given element is not constant; it varies depending on the element to which it is bound. Electronegativity generally increases across a period from left to right (say from lithium to fluorine) and decrease down a group (say from fluorine to astatine) in the periodic table. The attraction between the outer (or valence) electrons and the nucleus increases as the atomic radius decreases in a period. The electronegativity also increases.

On the same account electronegativity values decrease with the increase in atomic radii down a group. The trend is similar to that of ionization enthalpy. Non-metallic elements have strong tendency to gain electrons. Therefore, electronegativity is directly related to that non-metallic properties of elements. It can be further extended to say that the electronegativity is inversely related to the metallic properties of elements. Thus, the increase in electronegativities across a period is accompanied by an increase in non-metallic properties (or decrease in metallic properties) of elements. Similarly, the decrease in electronegativity down a group is accompanied by a decrease in non-metallic properties (or increase in metallic properties) of elements.

Trends in electronegativity across a period
As you go across a period the electronegativity increases. The chart shows electronegativities from sodium to chlorine - you have to ignore argon. It doesn't have electronegativity, because it doesn't form bonds.

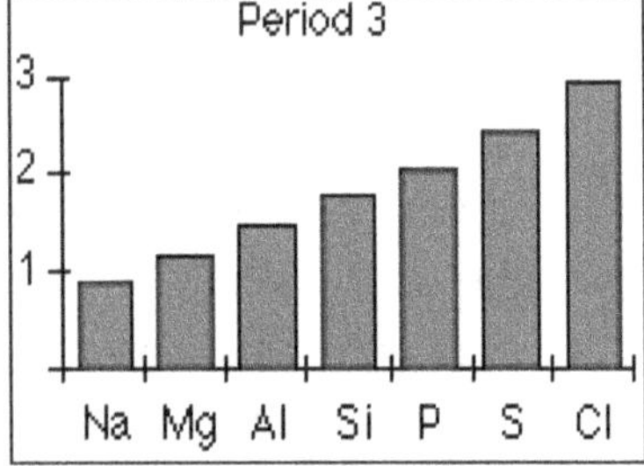

Trends in electronegativity down a group
As you go down a group, electronegativity decreases. (If it increases up to fluorine, it must decrease as you go down.) The chart shows the patterns of electronegativity in Groups 1 and 7.

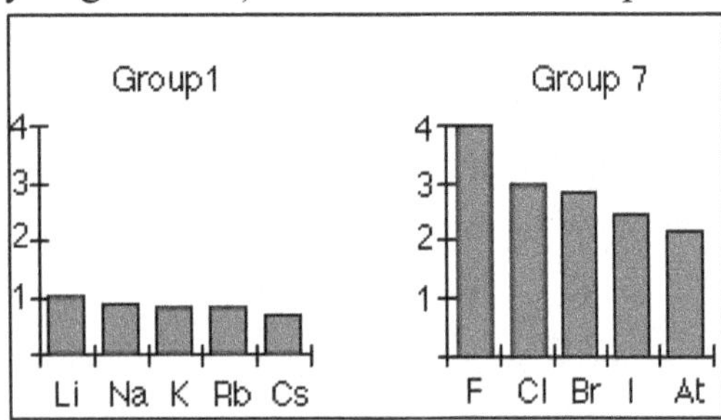

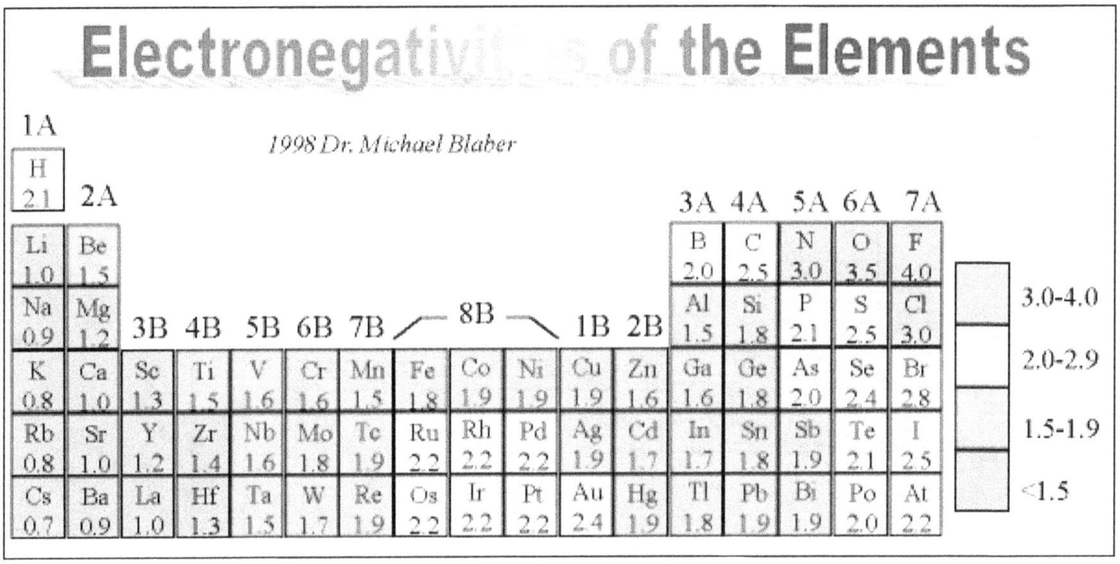

Electronegativities of the Elements

1998 Dr. Michael Blaber

1A	2A	3B	4B	5B	6B	7B	8B	8B	8B	1B	2B	3A	4A	5A	6A	7A
H 2.1																
Li 1.0	Be 1.5											B 2.0	C 2.5	N 3.0	O 3.5	F 4.0
Na 0.9	Mg 1.2											Al 1.5	Si 1.8	P 2.1	S 2.5	Cl 3.0
K 0.8	Ca 1.0	Sc 1.3	Ti 1.5	V 1.6	Cr 1.6	Mn 1.5	Fe 1.8	Co 1.9	Ni 1.9	Cu 1.9	Zn 1.6	Ga 1.6	Ge 1.8	As 2.0	Se 2.4	Br 2.8
Rb 0.8	Sr 1.0	Y 1.2	Zr 1.4	Nb 1.6	Mo 1.8	Tc 1.9	Ru 2.2	Rh 2.2	Pd 2.2	Ag 1.9	Cd 1.7	In 1.7	Sn 1.8	Sb 1.9	Te 2.1	I 2.5
Cs 0.7	Ba 0.9	La 1.0	Hf 1.3	Ta 1.5	W 1.7	Re 1.9	Os 2.2	Ir 2.2	Pt 2.2	Au 2.4	Hg 1.9	Tl 1.8	Pb 1.9	Bi 1.9	Po 2.0	At 2.2

Legend:
- 3.0-4.0
- 2.0-2.9
- 1.5-1.9
- <1.5

Fluorine is the ***most*** electronegative element (electronegativity = 4.0), the ***least*** electronegative is Cesium (notice that are at diagonal corners of the periodic chart)

General trends:
- **Electronegativity** is the relative tendency of a bonded atom to attract electrons to itself.
- Electro negativity ***increases from left to right*** along a period
- For the representative elements (*s* and *p* block) the electro negativity ***decreases as you go down*** a group
- An atom with extremely low electronegativity, like a Group I metal, is said to be **electropositive** since its tendency is to lose rather than to gain, or attract, electrons
- For the commonly encounted elements, the order in decreasing electronegativity is F > O > N ~ Cl > Br > C ~ S ~ I > P ~ H > Si

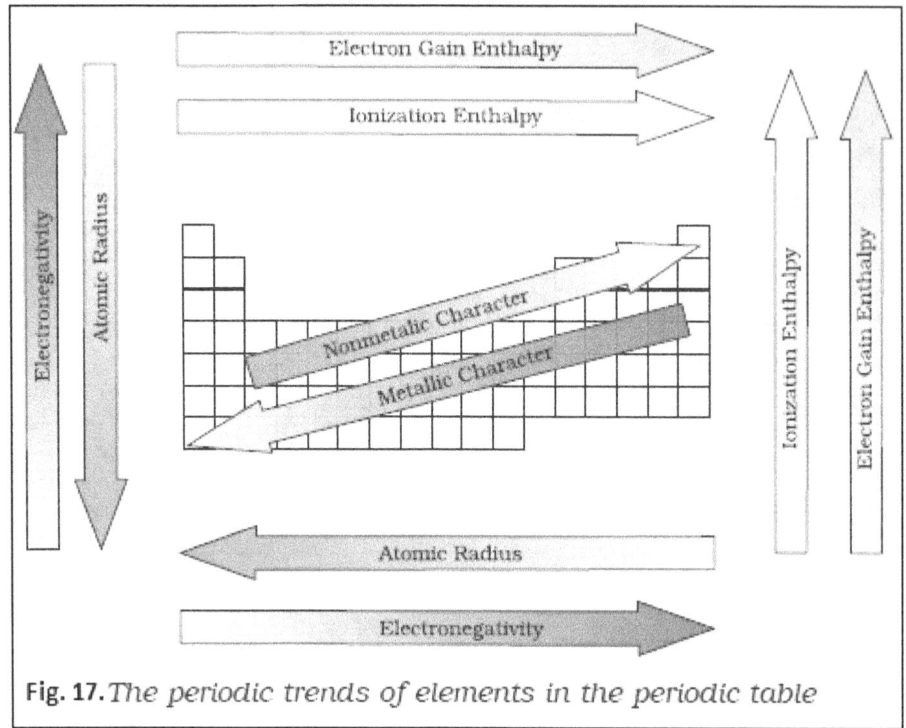

Fig. 17. *The periodic trends of elements in the periodic table*

2.7 PERIODIC TRENDS IN CHEMICAL PROPERTIES

It has been learnt that the atomic properties like atomic radius, ionization enthalpy, electron gain enthalpy and electronegativity exhibit periodicity. The periodicity of the valence state of the elements and anomalous properties of the second period elements from Li to F is discussed in the following section.

a) Periodicity of valence or oxidation states

The combining capacity of an element is termed as the valency of that element. Valency of an element is linked with the number of electrons present in the outer most energy shell of its atom. The electrons in the outer most shell (valence shell) are called valence electrons. The **valence (Oxidation state)** of an element may be defined as the **number of hydrogen or halogen atoms or twice the number of oxygen atoms which combine with one atom of the element.**

For example, the valency of carbon in CH₄ is 4, the valency of Nitrogen in NH₃ is 3, the valency of oxygen in H₂O is 2 and the valency of fluorine in HF is 1.

Nowadays the term oxidation state is frequently used for valence. Consider the two oxygen containing compounds: OF_2 and Na_2O. The order of electronegativity of the three elements involved in these compounds is F > O > Na. Each of the atoms of fluorine, with outer electronic configuration $2s^2 2p^5$, shares one electron with oxygen in the OF_2 molecule. Being highest electronegative element, fluorine is given oxidation state -1. Since there are two fluorine atoms in this molecule, oxygen with outer electronic configuration $2s^2 2p^4$ shares two electrons with fluorine atoms and thereby exhibits oxidation state +2. In Na_2O, oxygen being more electronegative accepts two electrons, one from each of the two sodium atoms and, thus, shows oxidation state -2. On the other hand sodium with electronic configuration $3s^1$ loses one electron to oxygen and is given oxidation state +1. Thus, the oxidation state of an element in a particular compound can be defined as the charge acquired by its atom on the basis of electronegative consideration from other atoms in the molecule. The variation in the valence of elements is shown in the Table 6 given below:

GROUP	1	2	13	14	15	16	17	18
NUMBER OF VALENCE ELECTRON	1	2	3	4	5	6	7	8
VALENCE	1	2	3	4	3,5	2,6	1,7	0,8

GROUP	1	2	13	14	15	16	17
Formula of hydride	LiH NaH KH	CaH$_2$	B$_2$H$_3$ AlH$_3$	CH$_4$ SiH$_4$ GeH$_4$ SnH$_4$	NH$_3$ PH$_3$ AsH$_3$ SbH$_3$	H$_2$O H$_2$S H$_2$Se H$_2$Te	HF HCl HBr HI

Table 6 Periodic Trends in Valence of Elements as shown by the Formulas

Formula of oxide	1	2	13	14	15	16	17
	Li$_2$O Na$_2$O K$_2$O	MgO CaO SrOM BaO	B$_2$O$_3$ AL$_2$O$_3$ Ga$_2$O$_3$ In$_2$O$_3$	CO$_2$ SiO$_2$ GeO$_2$ SnO$_2$ PbO$_2$	N$_2$O$_3$,N$_2$O$_5$ P$_4$O$_6$,P$_4$O$_{10}$ As$_2$O$_3$,As$_2$O$_5$ Sb$_2$O$_3$,Sb$_2$O$_5$ Bi$_2$O$_3$-	SO$_2$ SeO$_3$ TeO$_3$	- Cl$_2$O$_7$ - -

Periodic Trends in Valence of Elements as shown by the Formulas

b) Anomalous properties of second period elements

The first element of each of the groups 1 (lithium) and 2 (beryllium) and groups 13-17 (boron to fluorine) differs in many respects from the other members of their respective group. For example, lithium unlike other alkali metals, and beryllium unlike other alkaline earth metals, form compounds with pronounced covalent character; the other members of these groups predominantly form ionic compounds. On moving diagonally across the periodic table the elements show certain similarities. These similarities are more pronounced in the following pairs of elements.

Li Be B

Mg Al Si

Lithium is similar to Magnesium in many of its properties and beryllium is similar to aluminium. Therefore, diagonal relationship is the similarity between a pair of elements which are diagonally opposite in periodic table. So the two elements Li and Mg having diagonal relationship resemble very much in properties.

The anomalous behaviour is attributed to their small size, large charge/ radius ratio and high electronegativity of the elements. In addition, the first member of group has only four valence orbitals (2s and 2p) available for bonding, whereas the second member of the groups have nine valence orbitals (3s, 3p, 3d). As a consequence of this, the maximum covalency of the first member of each group is 4 (e.g., boron can only form $[BF_4]^-$, whereas the other members of the groups can expand their valence shell to accommodate more than four pairs of electrons e.g., aluminium forms $[AlF_6]^{3-}$). Furthermore, the first member of p-block elements displays greater ability to form $p_\pi - p_\pi$ multiple bonds to itself (e.g., $C = C$, $C \equiv C$, $N = N$, $N \equiv N$) and to other second period elements (e.g., $C = O$, $C = N$, $C \equiv N$, $N = O$) compared to subsequent members of the same group.

Periodic Trends and Chemical Reactivity

Periodic trends in certain fundamental properties such as atomic and ionic radii, ionization enthalpy, electron gain enthalpy and valence has been observed so far. We know by now that the periodicity is related to electronic configuration. That is, all chemical and physical properties are a manifestation of the electronic configuration of elements. We shall now try to explore relationships between these fundamental properties of elements with their chemical reactivity.

The atomic and ionic radii, as we know, generally decrease in a period from left to right. As a consequence, the ionization enthalpies generally and electron gain enthalpies become more negative across a period. In other words, the ionization enthalpy of the extreme left element in a period is the least and the electron gain enthalpy of the element on the extreme right is the highest negative (note : noble gases having completely filled shells have rather positive electron gain enthalpy values). This results into high chemical reactivity at the two extremes and the lowest in the centre. Thus, the maximum chemical reactivity at the extreme left (among alkali metals) is exhibited by the loss of an electron leading to the formation of a cation and at the extreme right (among halogens) shown by the gain of an electron forming an anion. This property can be related with the reducing and oxidizing behaviour of the elements which you will learn later. However, here it can be directly related to the metallic and non-metallic character of elements. Thus, the metallic character of an element, which is highest at the extremely left decreases and the non-metallic character increases while moving from left to right across the period. The chemical reactivity of an element can be best shown by its reactions with oxygen and halogens. Here, we shall consider the reaction of the elements with oxygen only. Elements on two extremes of a period easily combine with oxygen to form oxides. The normal oxide formed by the element on extreme left is the most basic (e.g., Na_2O), whereas that formed by the element on extreme right is the most acidic (e.g., Cl_2O_7). Oxides of elements in the centre are amphoteric (e.g., Al_2O_3, As_2O_3) or neutral (e.g., CO, NO, N_2O). Amphoteric oxides behave as acidic with bases and as basic with acids, whereas neutral oxides have no acidic or basic properties.

2.8 EXERCISE QUESTIONS
Exercise Questions: Type 1
1. Why does atomic radius decreases in a period?
2. Why does atomic radius increases in a group?
3. Radius of cation is less than that of the atom, Give reason.
4. Which element is expected to be most metallic in the periodic table?
5. Electron affinity of fluorine is less than that of chlorine, Give reason.
6. How reducing and oxidizing property varies in a period and a group.
7. How does nature of oxides varies in a period?
8. Which of the following ions has the largest ionic radius and which is the smallest Sc^{3+}, Cl^-, Ca^{2+}.
9. Arrange the following as diversified.
    i. Na, Rb, K, Mg in order of increasing atomic radius
    ii. Cl^-, S^{2-}, Ca^{2+}, Ar in order of increasing size.
10. What are iso-electronic species?
11. What is electronegativity?
12. Which is the most electropositive element?
13. Name the element having highest electron affinity.
14. Alkali metals have low ionization potential. Give reason.

Exercise Questions: Type 2
15. What are the merits of modern periodic table?
16. Explain the concept of periodicity.
17. Lanthanides and Actinides are placed separately at the bottom of the periodic table. Give reason.
18. Explain the following terms:
    (1) Covalent radius (2) Vaanderwaal's radius (3) Ionisation energy (4) Electron affinity
19. Mention the two factors on which electronegativity depends.
20. Discuss the variation of ionic size in isoelectronic species.
21. Discuss the variation of atomic radii/ionization energy across a period in periodic table.

Exercise Questions: Type 3
22. Explain the characteristic features of modern periodic table.
23. Explain the classification of elements into s, p, d and f blocks.
24. Give any two merits and demerits of modern periodic table.
25. Explain the factors affecting ionization energy.

Objective Type Questions
1. Which one of the following has lowest first I.E?
    a. Li   b. Na   c. Rb   d. Cs
2. The alkali metals are strong reducing agents because of
    a. Low IE b. Large ionic radii
    c. Low heat of sublimation   d. Potential value
3. Electron affinity is maximum in case of
    a. Cl   b. B   c. F   d. Na

4. Which of the following has least electronegativity?
   a. Li   b. F   c. Fr   d. Cl
5. The ionisation energy of nitrogen is greater than that of oxygen because of
   a. Nitrogen is present to left of oxygen in the same period.
   b. Less number of electrons in nitrogen
   c. Presence of half filled p-orbitals in nitrogen
   d. Less electronegetivity of N than O.
6. The biggest ion among the following is?
   a. Al^{3+} b. Ba^{2+} c. Mg^{2+} d. Na^{+}
7. The smallest cation among the following is
   a. Na^{+} b. Mg^{2+} c. Ca^{2+} d. Al^{3+}
8. The difference between the electronegativities of two elements represents
   a. Bond length   b. Size of anion
   c. Size of cation   d. nature of bond formed between them
9. Which one of the following groups of atoms or ions is not isoelectronic?
   a. He, H^{-}, Li^{+} b. Na^{+}, Mg^{2+}, Al^{3+}
   c. F^{-}, O^{2-}, N^{3-} d. K^{+}, Ca^{2+}, Ne
10. Electron affinity is negative in case of
    a. Li   b. Be   c. C   d. F
11. In which of the following energy is absorbed
    a. $F \rightarrow F^{-}$ b. $Cl \rightarrow Cl^{-}$ c. $O \rightarrow O^{2-}$ d. $H \rightarrow H^{-}$
12. The radius of isoelectronic species?
    a. Increase with increase in nuclear charge
    b. Decrease with increase in nuclear charge
    c. Same for all
    d. First increase and then decreases.
13. Which of the following electronic configuration corresponds to an element having lowest IE?
    a. $1s^2\, 2s^2\, 2p^6$ b. $1s^2\, 2s^2\, 2p^6\, 3s^1$
    c. $1s^2\, 2s^2\, 2p^5$ d. $1s^2\, 2s^2\, 2p^3$
14. Which of the following has the lowest IE?
    a. $3d^2$ b. $4s^1$ c. $2p^6$ d. $3p^6$
15. The electronic configuration of four elements are given below, which has the highest I.E
    a. $[Ne]\, 3s^2\, 3p^1$ b. $[Ne]\, 3s^2\, 3p^3$
    c. $[Ne]\, 3s^2\, 3p^2$ d. $[Ar]\, 3d^{10}\, 4s^2\, 4p^3$
16. The electronic configuration of oxide ion O^{2-} may be represented as
    a. $1s^2\, 2s^2\, 2p^4$ b. $1s^2\, 2s^2\, 2p^6$
    c. $1s^2\, 2s^2\, 2p^6\, 3s^2\, 3p^6$ d. $1s^2\, 2s^2\, 2p^6\, 3s^2\, 2p^4$

Chapter 3

CHEMICAL BONDING

3.1 INTRODUCTION

Compounds are formed from atoms reacting together. Now the question that might arise in your mind is that what is the underlying mechanism that holds the atoms together? There are millions of compounds that are formed from combination of different types of elements. For example Hydrogen, Nitrogen, Oxygen are basic elements, but in nature they are found to have a stable chemical formula H_2, N_2, and O_2. These are the stable molecules of the Hydrogen, Nitrogen and Oxygen gases. Sodium (Na) is a metal, Chlorine (Cl) is a non-metal, and salt that is NaCl is a compound.

There is a definite way in which atoms of one element interacts with other atoms. You will not have Sodium (Na) and Potassium (K) interacting with each other!! To understand how compounds are formed, we have to first know the types of bonds that can be forged between atoms. Electronic configuration and valence of the atoms play a crucial role in the types of bonds.

Most of the substances exist in the form of clusters or aggregates of atoms. Any such cluster, in which atoms are held together and which is electrically neutral is called a molecule. The attraction between atoms within a molecule is called a chemical bond.

Chemical bond is a sort of attraction, which holds the atoms together in a molecule.
When two atoms approach one another, a chemical bond is formed between them only if potential energy of the system consisting the two atoms undergoes a fall. If there is no fall in potential energy of the system, no bonding is possible.
The kind of chemical bonds formed by an atom depends upon its electronic configuration and position of the element in the periodic table [cause of chemical bonding].

<u>Type of bonds</u>
Atoms may attain a stable electronic configuration in three different ways by losing electrons, by gaining electrons, or by sharing electrons. Elements may be divided into:
1. Electropositive elements, whose atoms give up one or more electrons fairly readily
2. Electronegative elements, which will accept electrons
3. Elements which have little tendency to lose or gain electrons.
Electropositive element
      +   Ionic bond
Electronegative element

Electronegative element
      +   Covalent bond
Electronegative element

58

Electropositive element
+   Metallic bond
Electropositive element

[Cause of chemical combination]: Tendency of all elements to acquire noble gas configuration. Atoms of all elements tend to acquire the noble gas configuration which is of maximum stability and hence of sharing of electrons amongst themselves or by the process of transference of electrons from one atom to another. This is the fundamental cause of chemical combination in most cases]

3.2 IONIC BOND

An ionic bond is formed by the complete transference of one or more electrons from the outer energy shell of one atom to the outer energy shell of the other atom [valence shell]. In this way, both the atoms acquire the electronic configuration of the nearest noble gases.

An ionic bond may be formed by the transfer of one electron from the outermost energy shell of alkali atom to the outermost shell of halogen atom. The atom which lose electron will acquire positive charge, while the atom which gains electrons will acquire negative charge. These charged atoms are called ions and held together by electrostatic force of attraction in the molecule.

For example, consider the formation of sodium chloride [NaCl]. The electronic configuration Na is $1s^2$ $2s^2$ $2p^6$ $3s^1$ and that of chlorine is $1s^2$ $2s^2$ $2p^6$ $3s^2$ $3p^5$. The atomic number of Na is one more than that of Ne [$1s^2$ $2s^2$ $2p^6$] and that of chlorine is one less than that of Ar[$1s^2$ $2s^2$ $2p^6$ $3s^2$ $3p^6$]. Thus neutral sodium atom has a tendency to lose one electron from the outermost 3s orbital to acquire stable neon gas configuration. Sodium atom by losing one electron acquires positive charge [Na^+] (Fig. 18).

Na $\longrightarrow$ Na^+ + e^-
(neutral sodium atom)   (positive sodium ion)   (electron)

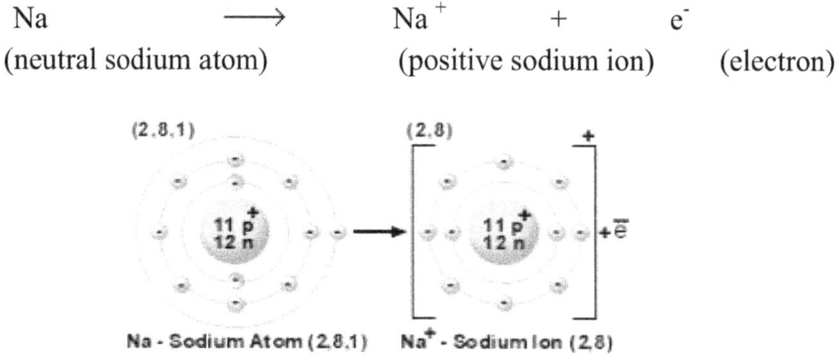

Fig. 18. Electron Transfer in sodium atom

Neutral chlorine atom has a tendency to gain one electron to acquire stable configuration of neon. Chlorine atom by gaining one electron acquires negative charge (Cl^-) (Fig. 19).

Cl   + e^- $\longrightarrow$ Cl^-
(neutral chlorine atom)   (electron)   (negative chloride ion)

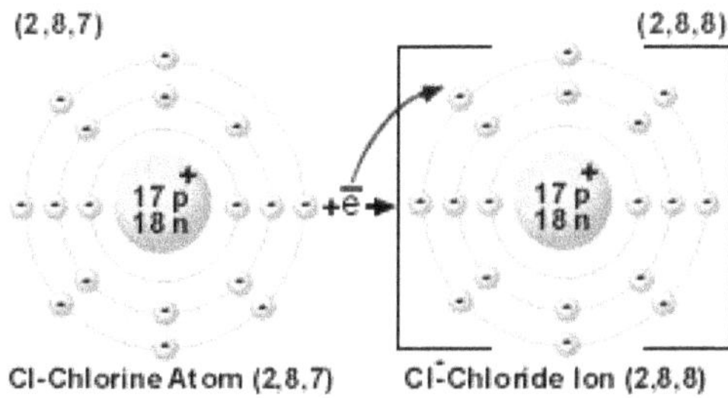

Fig. 19. Electron addition to Chlorine atom

The positive sodium ions and negative chloride ions are held by electrostatic force of attraction to form ionic solid sodium chloride [Na^+Cl^-] (Fig. 20).

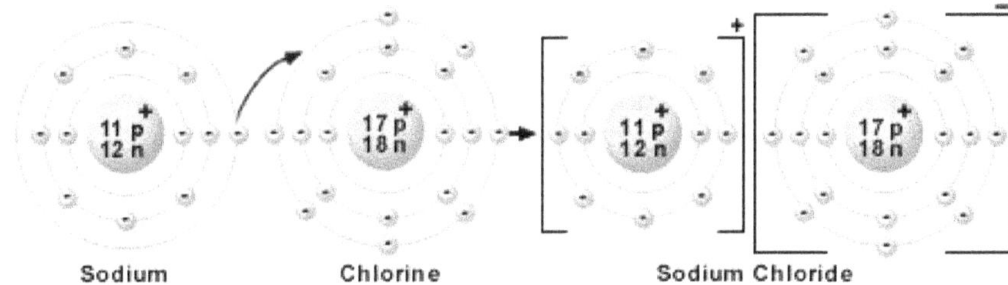

Fig. 20. Formation of sodium chloride with electron transfer

Hence sodium chloride contains sodium ions and chloride ions in the lattice. X-ray and spectroscopic techniques reveal the presence of ions in ionic crystals.

Similarly calcium chloride is formed by the combination of calcium and fluoride ions. Calcium transfers two electrons from its outermost shell to two fluoride ions.

In the similar way, we can explain the formation of ionic bonds in a large number of compounds e.g., LiF, Licl, KCl, KOH, CaF_2, MgO, $MgCl_2$, Na_2S, MgS etc.

Conditions for the formation of ionic bonds
An ionic compound contains a metal ion (cation) and a non-metal ion (anion). The following factors play important role in the formation of ionic bond between metal and non-metal ions.
1. **Low ionization energy for metal atom**: In the formation of ionic bond, metal atom must have a low ionisation energy so that it easily loses electron

   (Li > Na > K > Rb > Cs)

2. **High electron affinity for non-metal atom**: The non-metal atom must have a high electron affinity i.e., capacity to hold extra electrons (F > 0 > N > C) greater the electron affinity of an atom higher is the tendency to form anions.

3. **High Lattice energy**: The electrostatic attraction between charged ions in the crystal i.e., lattice energy should be high. The magnitude of the lattice energy depends upon

(a) Size of the ions: Smaller the size of the ions larger will be the magnitude of lattice energy. For example, $NaCl \rightarrow 782$ kJ/mol.

(b) Charge on the ions: Larger the magnitude of charge on the ions greater will be the attractive force and consequently higher is the value of lattice energy. For example, $LiF \rightarrow 1.37$ kJ/mol

4. Large differences of electro negativities between the electro positive element (metal) and electro negative element (non-metal).

Properties of ionic compounds
1. Ionic compounds are usually crystalline solids that are hard and brittle.
2. Ionic substances are freely soluble in polar solvents like water and are insoluble in solvents like benzene, CS_2. CCl_4 etc., [H_2O has high dielectric constant]
3. Ionic compounds donor conduct electricity in solid state but act as good conductors in fused state.
4. The chemical reactions between ionic compounds in solutions are very fast.
5. They do not exhibit space isomerism due to non-directional nature of ionic bond.

Lattice Energy
The lattice energy (U) of crystals is the energy evolved when one mole of ionic crystal is formed from gaseous ions.
Example:

$$Na^+(g) + Cl^{-1}(g) \longrightarrow NaCl \ (crystal) \quad \Delta H = -758.7 \ kJ \ mol^{-1}$$

$$Li^+(g) + F^-(g) \longrightarrow LiF \ (crystal) \quad \Delta H = -1004 \ kJ \ mol^{-1}$$

- Lattice energy increases with decrease of inter-ionic distance.

  Lattice energy of LiF ($r = 2.01$ Å) is greater than that of MgO ($r = 2.10$ Å)
- Crystals with high lattice energy usually possess high melting points.
- Lattice energies cannot be measured directly, but experimental values are obtained from thermodynamic data using the Born-Haber cycle.

3.3 COVALENT BOND
A covalent bond is formed by the common sharing of an electron pair between two atoms each of which contribute one electron to the electron pair. The way of combinations of two atoms involving mutual sharing of electrons is called covalency. The shared electrons contribute towards the stability of both the atoms.

The compounds formed as a result of covelency are called covalent compounds or non-polar compounds as the electron pairs are shared equally by the consultant atoms

For example, H_2 molecule contains a single covalent bond. Each H atom contributes one electron towards formation of a covalent and attain complete outer shell configuration i.e., $1s^2$. $\quad$ H . + .H $\rightarrow$ H:H
H-H

Similarlycovalentbond formation in HF, H_2O, NH_3 and CH_4 can be accounted as follows:

$$H\cdot + \cdot\ddot{\underset{..}{F}}: \longrightarrow H:\ddot{\underset{..}{F}}:$$

$$2H\cdot + \cdot\ddot{\underset{.}{O}}: \longrightarrow H:\overset{..}{\underset{H}{O}}:$$

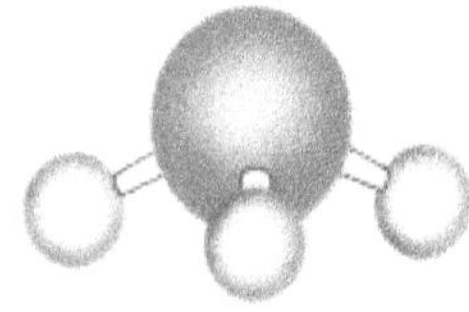

$$3H\cdot + \cdot\overset{..}{\underset{.}{N}}\cdot \longrightarrow H:\overset{..}{\underset{H}{N}}:H$$

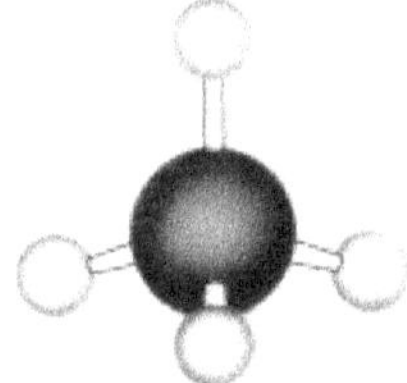

$$4H\cdot + \cdot\overset{.}{\underset{.}{C}}\cdot \longrightarrow \overset{H}{\underset{H}{H:C:H}}$$

The number of covalent bonds formed between atoms of any two elements depends on their outermost electronic configuration i.e., Number of electrons required to attain the nearest noble gas configuration.

The covalent bond formed by sharing of one pair of electrons is called single bond, by sharing of two pairs of electrons is called a double bond and by sharing of three pairs of electrons is known as triple bond.

Examples:

In the oxygen molecule, each atom has six valence electrons so for each to achieve an octet configuration; two pairs of electrons must be shared making a double bond between the atoms.

$$:\ddot{\underset{.}{O}}\cdot + \cdot\ddot{\underset{.}{O}}: \longrightarrow :\ddot{\underset{..}{O}}::\ddot{\underset{..}{O}}: \text{ or } \ddot{\underset{..}{O}} = \ddot{\underset{..}{O}}$$

Similarly, the N_2 molecules have a triple bond, involving three shared electron pairs.

$$\cdot\ddot{\underset{.}{N}}\cdot + \cdot\ddot{\underset{.}{N}}\cdot \longrightarrow :N:::N: \text{ or } :N \equiv N:$$

Properties of covalent compounds

1. Covalent compounds are usually gases and liquids under normal conditions.

2. These compounds are generally slightly soluble in water, but readily soluble in organic solvents such as benzene, CCl_4 etc.
3. These compounds exist as molecules and so do not conduct electricity either in fused state or in dissolved state.
4. The covalent bond is rigid and directional and hence there is a possibility of position and stereo isomerism.

Factors favouring covalent bonds

The formation of covalent bond is favoured under the following conditions:
(i) The combining atoms should have small difference in electro negativity.
(ii) The combining atoms should attain octet structure by sharing one or more electrons.
(iii) Atoms should have small size.
(iv) Atoms should have high electronic affinity.
(v) Atomic orbitals should contain unpaired electron.

3.4 VALENCE BOND THEORY

The important postulates of valence bond theory are:
(i) Covalent bond is formed by overlapping of half filled atomic orbital of one atom with half filled atomic orbital of the other atom involved in the bond formation.
(ii) A chemical bond is formed only when two electrons of opposite spins are paired.
(iii) The stronger bond will be formed between the orbital of two atoms that overlap to the maximum extent.
(iv) The direction of the bond formation will be in that direction in which orbitals are concentrated.
(v) The extent of overlapping of orbitals follows the order

      s - s overlap > s - p overlap > p - p overlap

Covalent bonds are of two types:
(i) Sigma (σ) bond and [A sigma bond is formed by the linear or end-to-end overlap of orbitals]
(ii) Pi [π] bond [A pi bond is formed by the lateral or sideways overlap of orbitals]
(i) Sigma (σ) bond

The covalent bond formed by the overlapping of two half filled orbitals along their axes (end-to-end overlap) is called a sigma bond.

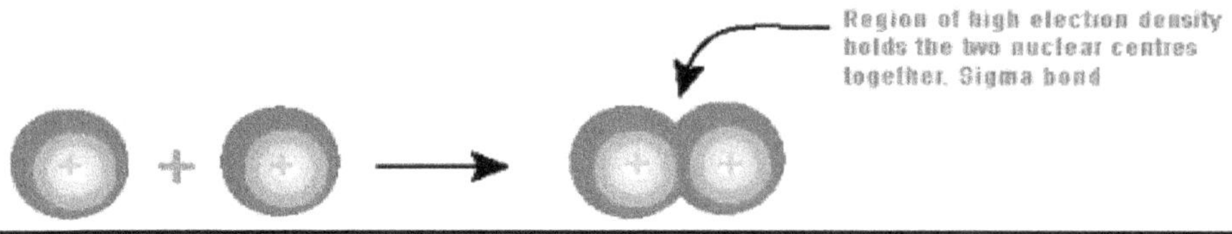

Types of overlap

a) s - s overlap: i.e., overlap of s-orbital of one atom by s-orbital of the other.

Example: In the formation of stable H_2 molecule two spherical s-orbital of each hydrogen atom overlap with each other to form a covalent bond. This bond is strong and is known as S-S bond

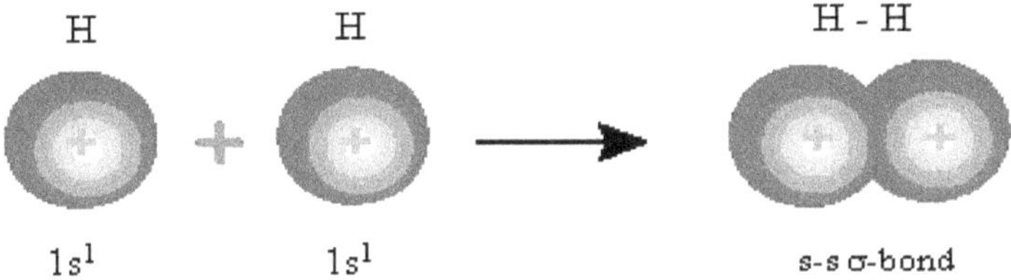

The electron density concentrates along the line joining the two bonded atoms, and it is symmetrical about the line joining two nuclei.

c) **s-p overlap:** i.e., overlap of s-orbital of one atom by p-orbital of the other.

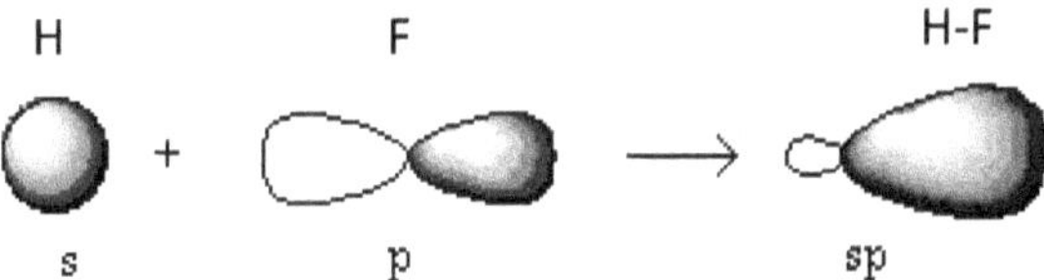

Example: HF molecule is formed by the overlapping of the half filled 1s orbital of hydrogen with half filled 2p orbital of fluorine along the axis.

c) p-p overlap [Head on overlap] end to end overlap of two p-orbitals

Example: In the formation of F_2 molecules, the half filled 2p orbital of a fluorine atom overlaps with the half filled 2p orbital of the other fluorine atom along the axis, to form a p-p sigma bond.

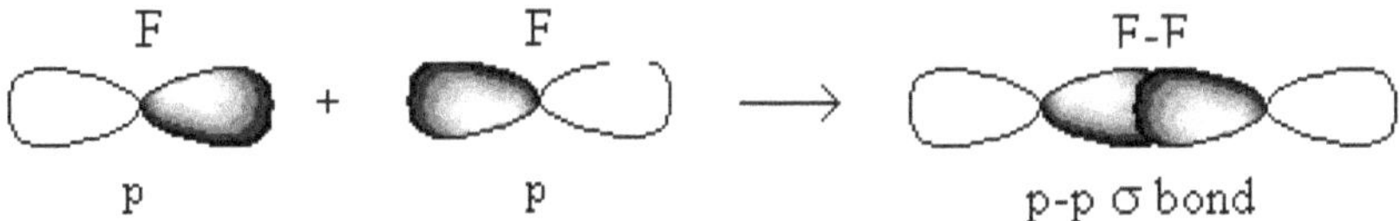

(ii) Pi (π) bond

i) A covalent bond formed between two atoms by the overlap of p-orbitals along a line perpendicular to the molecular axis is called a π -bond.

Electron cloud of a π (pi) bond is unsymmetrical and is concentrated above and below the plane of the sigma bond.

Sideways overlapping is less efficient than one along the axis. So, the π bond is weaker than the σ bond. Example: oxygen molecule.

i) Oxygen molecule
Consider the electronic configuration of Oxygen atom.
Oxygen needs two electrons to complete its octet. So it has the tendency to share two of its unpaired electrons with other oxygen atom and two bonds exist between two oxygen atoms.

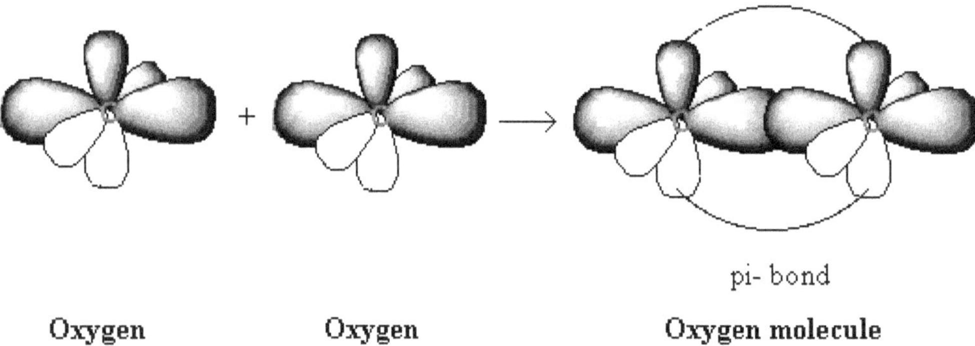

Fig. 21. Formation of oxygen molecule

In O_2 molecule $2P_x$, $2P_y$ orbitals contain unpaired electrons. $2P_x$ orbitals of both oxygen atoms undergoes end to end overlap resulting in the formation of σ bond. $2P_y$ orbitals of both atoms parallelly under sideways overlap resulting in the formation of π bond (Fig. 21).

3. 5 HYBRIDISATION AND BONDING IN SIMPLE MOLECULES
Hybridisation is the process of mixing of pure orbitals of almost same energy to give a set of new equivalent orbitals called hybrid orbitals.

Number of Groups Attached to a Central Atom	Description and 3-Dimensional Shape
Two Groups...sp	2 groups = sp hybridization 180 degree bond angle linear electron-pair geometry
Three Groups...sp²	3 groups = sp² hybridization 120 degree bond angles trigonal planar electron-pair geometry
Four Groups...sp³	4 groups = sp³ hybridization 109.5 degree bond angles tetrahedral electron-pair geometry

Bonding in methane

The electronic configuration of carbon in its ground or atomic state is

$$C_{\text{Ground State}} = 1s^2\, 2s^2\, 2p_x^1\, 2p_y^1,\, 2p_z^0$$

Carbon gets excited to the empty P_z orbital. The energy required to promote electron is supplied in the form of heat or light.

$$C_{\text{Excited State}} = 1s^2\, 2s^1\, 2p_x^1\, 2p_y^1\, 2p_z^1$$

The 2s orbital and three 2p orbital (p_x, p_y and p_z) in the excited state are mixed or hybridized to give four new equivalent orbitals. These new orbitals are known as sp³ hybrid orbitals or simple sp³ orbitals because they are formed by the interaction of one s and three p orbitals. The electronic configuration of the carbon atom in it's sp³ hybridized state. $C_{\text{Hybrid State}} = 1s^2\, 2(sp^3)^1\, 2(sp^3)^1\, 2(sp^3)^1\, 2(SP^3)^1$

Each sp³ orbital contain one electron. Each sp³ orbital has a large lobe and a small lobe. The four sp³ orbitals are arranged in such a way that their oxes are directed towards the corners of a regular tetrahedron. In methane, carbon forms single covalent bonds with four hydrogen atoms. Here 4 sp³ orbitals of carbon overlap with s-orbitals of 4 hydrogen atoms leading to the formation of four C-H bonds which are directed towards the corners of a regular tetrahedron as shown

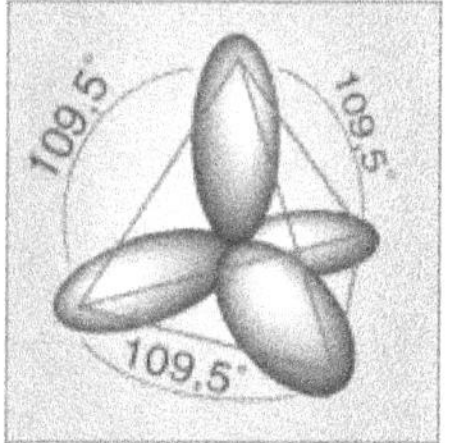

below. The covalent bonds formed by the overlap of sp^3 orbitals and s orbitals are sigma (σ) bonds because the electron density in each bond is symmetrical about the line joining the centres of two bonded atoms. Thus all C-H bonds in methane are sigma bonds as illustrated in Fig. 22.

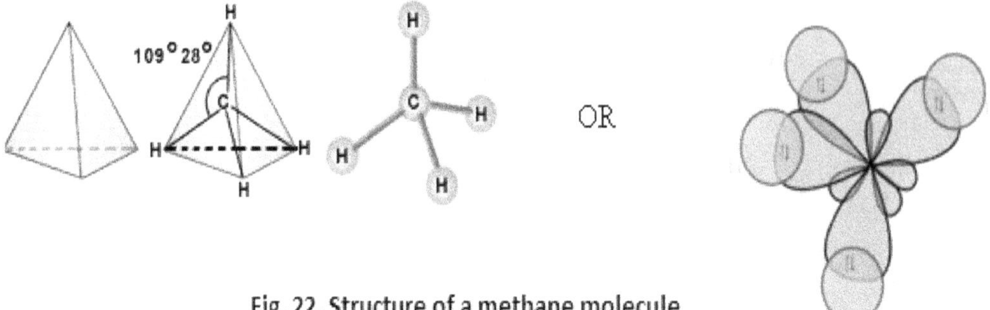

Fig. 22. Structure of a methane molecule

Bonding in ethylene [C$_2$ H$_4$]

The electronic configurations of carbon in ground and exited states are

$$C_{Ground\ State} = 1s^2[2s^2\ 2p_x^{'}\ 2p_y^{'}\ 2p_z^{o}]$$

$$C_{Excited\ State} = 1s^2[2s^{'}\ 2p_x^{'}\ 2p_y^{'}\ 2p_z^{'}]$$

Carbon atoms in ethylene, undergoes sp^2 hybridisation i.e., 2s orbital and just two of the three 2p orbitals are mixed or hybridized to give three new equivalent orbitals. These new orbitals are refered to as sp^2 orbitals, because they are formed by interaction of one s and two p orbitals. Third 2p$_z$ orbital is left unhybridised. The electronic configuration of the carbon atom in its sp^2 hybridised state is

$$C_{Hybrid\ State} = 1s^2[2(sp^2)^{'}\ 2(sp^2)^{'}\ 2(sp^2)^{'}\ 2p_z^{'}]$$

The sp^2 orbitals obtained above are identical but differ only in their orientation in space with respect to each other. The three sp^2 orbitals lie in the same plane with their axes directed towards the corner of an equilateral triangle. The angle between any pair of orbitals is thus 120°.The unlybridised 2p$_z$ orbital is oriented along an axis perpendicular to the plane sp^2 orbitals, with each lobe above and below the plane of the sp^2 orbitals (Fig. 23).

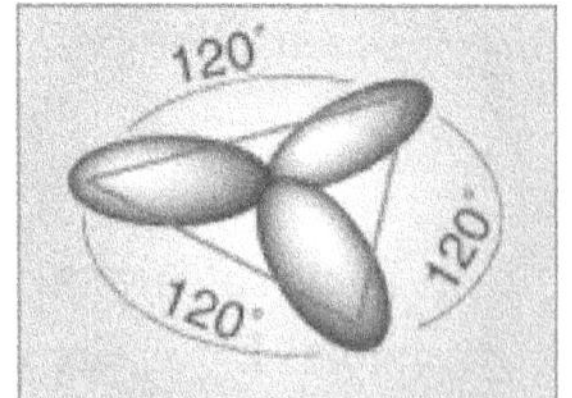

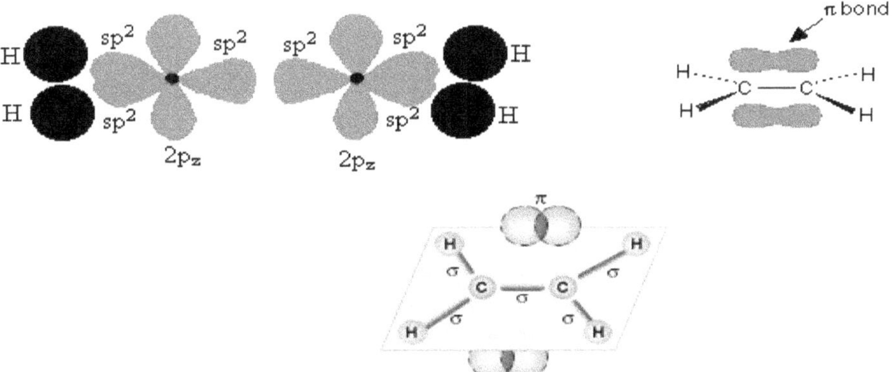

Fig. 23. Structure of ethylene molecule

In ethylene there are four C-H single covalent bonds and one C-C double bond. Each c-H bond is a sigma (σ) bond and results from the overlap of an sp^2 orbital from carbons with is orbital of hydrogen. One of the two bonds in the double bond is also a σ bond. Other bond is a π bond obtained by overlapping of unhybridized $2p_z$ orbital sideways.

Bonding in acetylene [$C_2 H_2$]

The electronic configuration in ground and excited states are

$$C_{Ground\ State} = 1s^2 [2s^2\ 2p_x^1\ 2p_y^1\ 2p_z^0]$$

$$C_{Excited\ State} = 1s^2 [2s^1\ 2p_x^1\ 2p_y^1\ 2p_z^1]$$

Carbon atoms in acetylene undergoes sp hybridization i.e., 2s orbital and just one of the three 2p orbitals are mixed or hybridised to give two new equivalent orbitals. These two new orbitals are called sp orbitals because they are formed by interaction of one s-orbital and one p-orbital. The other two 2p orbitals (p_y and p_z) are left unhybridised. The electronic configuration of the carbon atom in its sp hybridised state is

$$C_{Hybrid\ State} = 1s^2 [2(sp)^{2(sp^1)}\ 2p_y^1\ 2p_z^1]$$

sp orbitals contains unpaired electron and lie in a straight line. The angle between two sp orbitals is thus 180°. The unhybridized p_y and p_z orbitals are at right angles to the line of the sp hybrid orbitals orientation of sp hybrid orbitals is shown below:

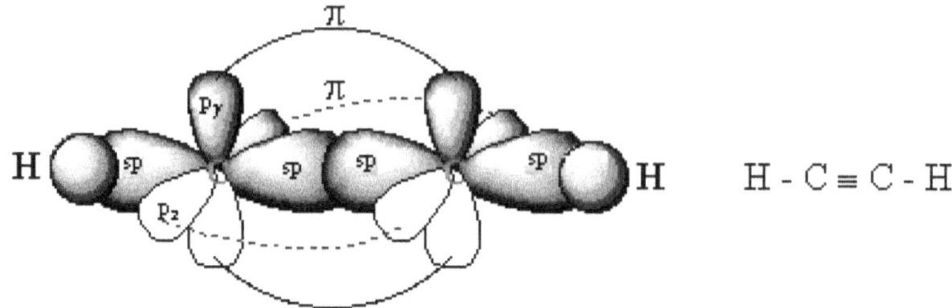

Sideways overlap of two unhybridized p orbitals p_y and p_z bonds formation or shown in Fig. 25,

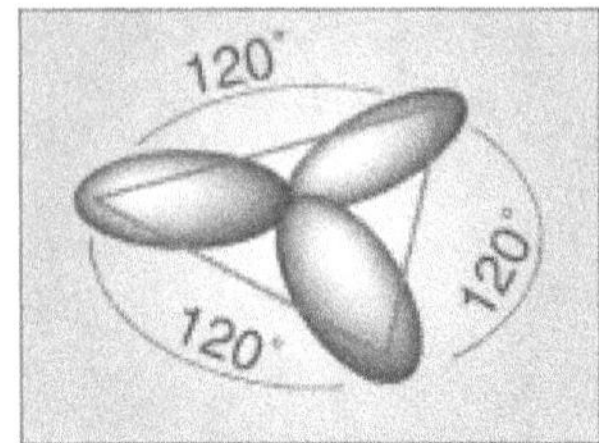

Fig. 25. Structure of ethylene molecule

Acetelyne molecule contains two C-H σ bonds as a result of overlap of an sp orbital from carbon and 1s orbital from hydrogen. Two carbon atoms in acetylene are linked together by three covalent bonds, of which one is σ and the other two are π -bonds. C-C triple bond in acetylene is represented by three equivalent line. One line represents a σ bond and the other two the π bonds.

Bonding in Boron trifluoride [BF₃]

The electronic configuration of Boron in its ground state is

$$\mathbf{B}_{\text{Ground State}} = 1s^2\, 2s^2\, 2p^1$$

Boron gets excited by the promotion of electron from 2s orbital to 2p orbital

$$\mathbf{B}_{\text{Exited State}} = 1s^2\, 2s^1\, 2p_x^1\, 2p_y^1$$

Boron atom in BF_3 undergo sp^2 hybridisation i.e., 2s orbital, $2p_x$ and $2p_y$ orbitals are mixed to give three new equivalent orbitals. These three hybrid orbitals of Boron overlap with $2p_z$ orbital of Fluorine to give trigonal planar Boron trifluoride molecule.

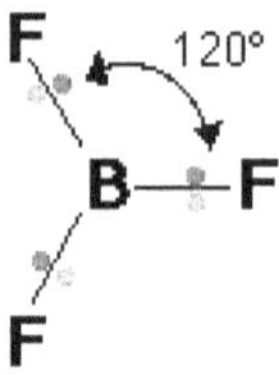

3.6 VALENCE SHELL ELECTRON PAIR REPULSION THEORY (VSEPR) THEORY

Sidgwick-Powell proposed VSEPR theory to predict and explain molecular shapes, based on electron pair repulsion in the valence shell. This theory was later developed by Gillespie and Nyholm in 1957.

Postulates of VSEPR theory:

1. The shape of the molecule depends on the number of electron pairs (both bond pairs and lone pairs) present in the valence shell of the central atom.

2. The electron pairs orient in the space around the central atom so as to have minimum repulsion among them.

3. A lone pair of electrons takes up more space around the central atom than a bond pair, since the lone pair is attracted to one nucleus while the bond pair is shared by two nuclei.

The order of repulsion among the electron pairs is:

lonepair - lonepair > lone pair - bond pair > bond pair - bond pair.

4. The presence of lone pairs at the central atom causes distortion in molecular shapes and decrease in bond angles.

5. Double bonds cause more repulsion than single bonds, and triple bonds cause more repulsion than a double bond.

Number of electron pairs	Shape of molecule	Bond angles
2	Linear	180°
3	Trigonal planar	120°
4	Tetrahedron	$109^\circ 28'$
5	Trigonal bypiramid	120° and 90°
6	Octahedron	90°
7	Pentagonal bipyramid	72° and 90°

Application of VSEPR Theory

1. Ammonia molecule (NH_3): Nitrogen atom in Ammonia involve sp^3 hybridisation. One of the four sp^3 hybrid orbitals contain one lone pair of

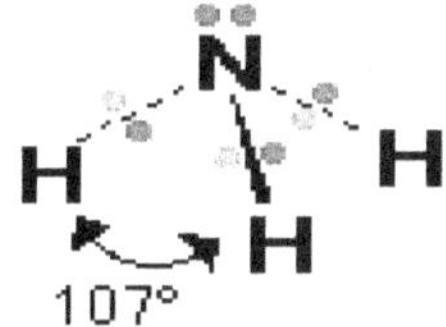

electrons. Other three sp^3 orbitals overlap with 3 hydrogen s orbitals to form 3 N-H σ bonds.

In ammonia molecule, the central atom nitrogen posses four electron pairs [one lone pair and 3 bond pairs]. These four electron pairs arrangement results in tetrahedral structure. Because of the repulsive interactions of lone pair with bond pair, the bond angle H-N-H is reduced from theoretical tetrahedral angle of $109° 28'$ to $107°48'$. Thus NH_3 molecular possess pyramidal shape.

2. **Water molecule**: In water molecule, oxygen involves sp^3 hybridisation. It contains 2 lone pairs of electrons and 2 bond pairs. The shape of the water molecule is based on a tetrahedron with two corners occupied by bond pairs and the other two corners occupied by lone pairs. Because of the greater repulsions between two lone pairs and lone pair-bond pair, the H-O-H bond angle.In water, bond angle reduces from $109° 28'$ to $104° 27'$. Thus H_2O molecule possesses bent or   v-shape.

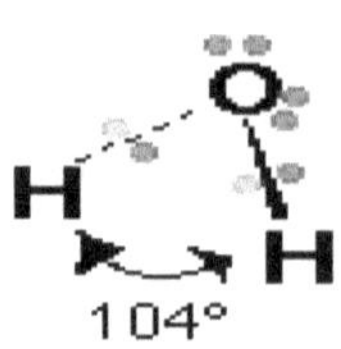

Molecule or ion	Effective number of electron pairs	Geometry around central atom	Shape
SO_2	3 (sulfur can 'expand' its octet)	Trigonal planar	
O_3	3 (double bonds count as 1 pair)	Trigonal planar	
IF_7	7	Pentagonal bipyramidal	
SF_4	5	Trigonal bipyramidal (distorted)	
NO_3^-	3	Trigonal planar	

3.7 POLAR AND NON-POLAR BONDS

A covalent bond formed between two identical or similar atoms is said to be non-polar covalent bond, but if formed between two dissimilar atoms, the bond formed is said to be polar covalent bond. In a non polar molecule, the shared electron pair is attracted equally by both atoms and lies exactly midway between them as in hydrogen molecule H:H,

Ex: F_2, Cl_2, Br_2, N_2, O_2, etc.

In a polar molecule covalent bond is formed between two dissimilar atoms, one of the atoms generally has a higher affinity for the electrons. The electron pair is passed closer to that atom which has more electro negativity

Example: (i) **Hydrogen chloride** (HCl) molecule, the electron pair spends more time near chlorine than near hydrogen atom. The unsymmetrical distribution of electrons leads to charge separation, i.e., development of partial negative charge near chlorine end and partial positive charge near hydrogen end.

This is represented as $\overset{\delta+\quad\ \delta-}{\text{H} - \text{Cl}}$

The polar covalent bond, therefore, has partial ionic character. The two opposite charges at the ends are called electrical poles and the molecules are called dipoles.

(ii) **Water** (H_2O) : Water molecule contains highly electro negative oxygen. Hence O-H bonds in water molecule are polar and possess moment. Thus water molecular has a net dipole moment of 1.85 D.

(iii) **Methane**: In methane, C-H bond also exhibits polarity due to the difference in the electro negativity between carbon and hydrogen. However, methane molecule has zero dipole moment due to highly symmetric tetrahedral structure. In this molecule, bond moments of each C-H bond gets cancelled.

Hydration of Ions

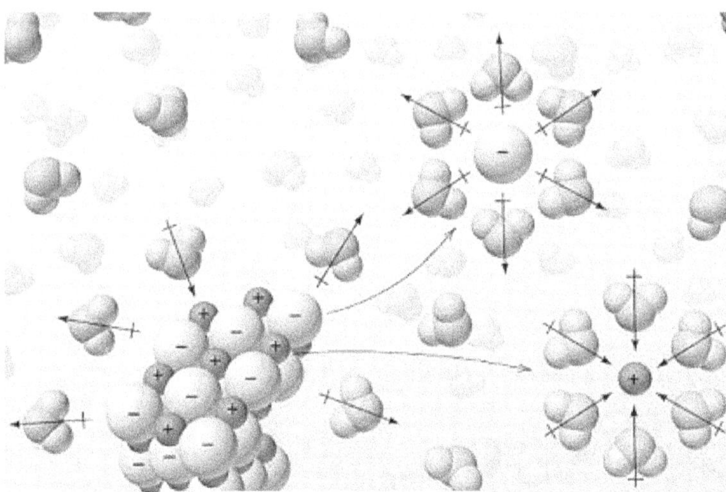

Fig. 26. Hydration of ions

All ionic solids dissolve in water. Water is a polar molecule. Water molecule has partial negative charge on oxygen atom and partial positive charge on Hydrogen atoms. When an ionic solid is added to water, negative part of water molecular attracts positive ions and positive part attracts negative ions. Every cation and Anion of ionic solid is surrounded by water molecules (Fig. 26).

Thus the ions are said to be hydrated and possess hydration sphere. This process where ions of ionic solid get hydrated is called hydration.

Since Li^+ is very small, it is heavily hydrated. The ability of the cation to get hydrated, decreases with size. This it follows the order;

$$Li^+ > Na^+ > K^+ > R_b^+ > Cs^+$$

3.8 DIPOLE MOMENT (μ)

Molecules containing different types of atoms usually exhibits polarity. These polar molecules possess dipole moment due to charge separation.

Dipole moment is defined as the product of the electric charge (e) and the distance between positive and negative centers of a polar molecule.

i.e., Dipole moment $\mu = e \times d$

The unit of dipole moment is debye (D)

$1D = 3.338 \times 10^{-30}$ coulomb meter.

Dipole moment of a molecule is a vector quantity. Therefore, an arrow pointing towards the electro negative atom represents it

Examples:

(1) Carbon dioxide (CO_2)

Symmetrical molecules like CO_2 have zero dipole moments. The C=0 bond in CO_2 has permanent dipole moment. CO_2 has linear structure.

$\mu = 0$	$\leftarrow \quad \rightarrow$ $O = C = O$

The two dipole moments neutralize with in the molecule and hence the net dipole moment for CO_2 is zero.

2. Water (H_2O)

Water is a bent or V shaped molecule. It has a net dipole moment of 1.84 D.

$\mu = 1.85\,D$	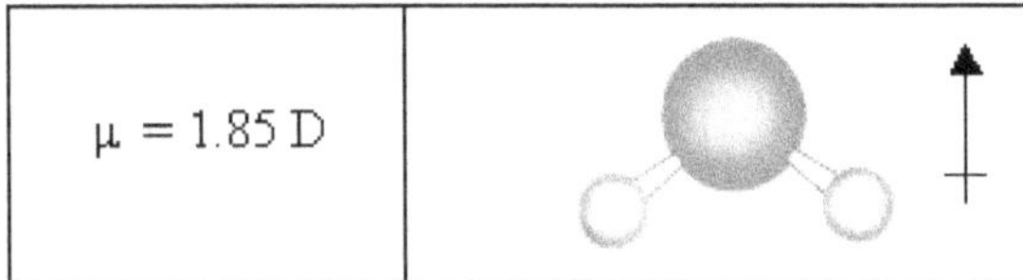

3. Ammonia (NH$_3$)

Ammonia molecule is pyramidal in its structure. Every N-H bond has moment. The net dipole moment of NH$_3$ is 1.49 D. The lower dipole moment value in NH$_3$ is due to low electro negativity of nitrogen compared to oxygen.

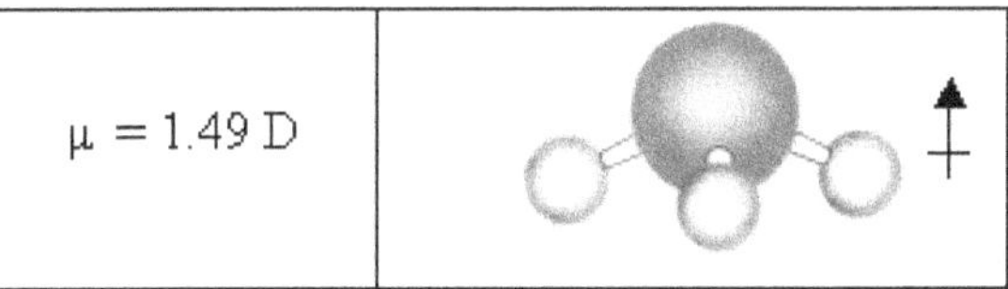

4. Boron trifluoride:

Boron trifluoride has trigonal planar structure. Since the molecule is symmetric. The dipole moment of BF$_3$ is zero.

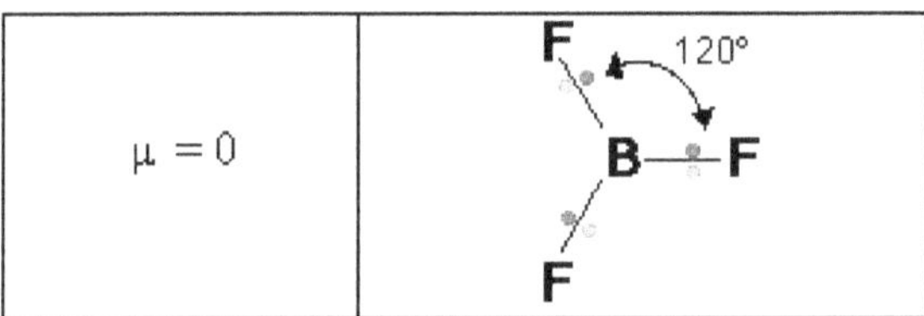

The degree of polarity of a bond or a molecule is usually expressed in terms of dipole moment which is defined as the product of the magnitude of the charges and the distance between them i.e., the bond length.

3.9 Coordinate bond (Dative bond)

In 1921, Perkins proposed the concept of coordinate covalent bond. According to his theory, Co-ordinate or Dative bond is formed by sharing of an electron pair between two atoms, but one of the two atoms contribute electron pair towards bonding.

An atom or group of atoms which contributes or donates electron pair to form coordinate bond is said to be donar atom or ligand.

Coordinate bond is represented as $A \xrightarrow{\bullet\bullet} B$ where A is a donor atom and B is a acceptor atom.

For example, in the formation of hydronium ion, oxygen atom in water molecule is the donar where as the hydrogen ion is the acceptor.

This type of bond is represented by an arrow pointing away from a ligand.

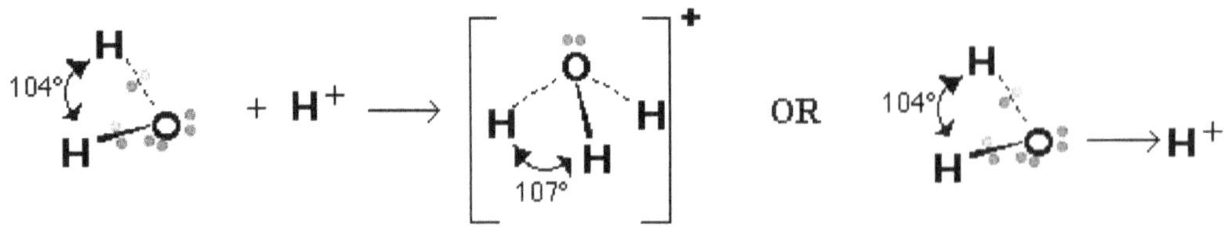

NH_3 has pyramidal structure. It has one lone pair. N atom can donate this lone pair of electrons to electron deficient H^+ ion to form co-ordinate molecular ion NH_4^+.

$$NH_3 + H^+ \longrightarrow H_3N \rightarrow H^+ \quad \text{or} \quad NH_4$$

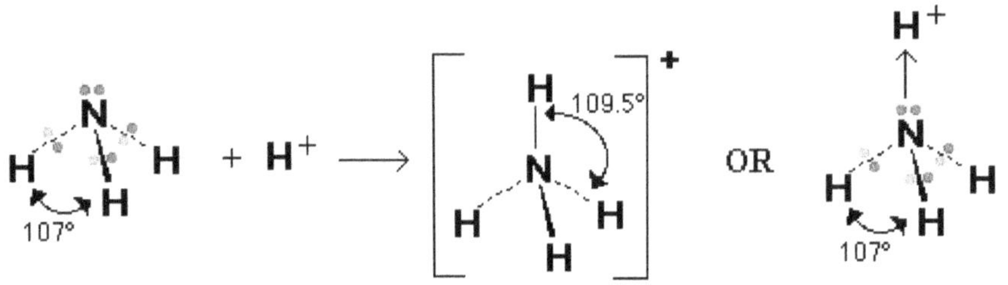

Ammonia forms an addition compound with boron trifluoride by the donation of an electron pair from nitrogen to boron.

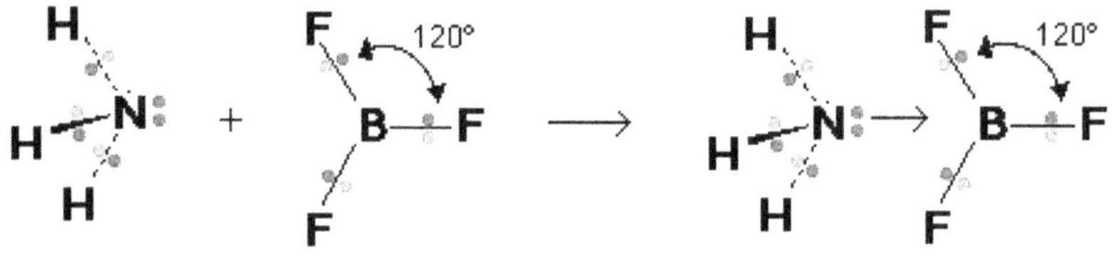

Characteristics of Co-ordinate compounds

a. They are usually insoluble in water, but soluble in organic solvents.

b. They usually donot conduct electricity

c. The M.P and B.Ps of coordinate compounds are higher than those of covalent compounds but lower than those of electrovalent compound.

3.10 Hydrogen bond

In certain molecules, when hydrogen lies between two strong electronegative atoms the hydrogen atom shows a unique property to form a bond or bridge between them with one electronegative atom it forms a normal covalent linkage, while with other it forms a weak linkage called hydrogen bond.

The attractive force while binds hydrogen of one molecule with electronegative atom of another molecule of the same substance is known as hydrogen bond.

For example, in case of HF the hydrogen atom of one molecule tends to attract fluorine atom in a neighbouring molecule and this attractive force forms a linkage between the hydrogen and electronegative fluorine atom. This is called hydrogen bonding which results in the binding of a large number of molecules to give an associated molecule, e.g, $(HF)_n$

H-F ….. H-F ……H-F ……H-F

Other example: H_2O, NH_3, HCL, HBr etc.

- The energies of hydrogen bonds are in the range of 2 to 10 K cal/mole as compared to the range of 30-130 K cal/mole for ordinary chemical bonds.

- Hydrogen bond occurs in compounds like water (H_2O), alcohol (R-OH), carboxylic acids [R-COOH], amines [R-NH_2] etc.

- Greater the electro negativity and smaller the size of the anion [F, O, N] the stronger is the hydrogen bond.

H-F   …. H   > H-O….H >   H-N …. H

10 Kcal/mol   7 Kcal/mol   20 Kcal/mol

- There are two types of hydrogen bonding:

i. Inter molecular hydrogen bonding

Inter molecular hydrogen bond is formed between two different types of molecules of the same or different substances. Example: (i) Hydrogen bonding in HF molecule (ii) Hydrogen bond between water molecules (iii) Hydrogen bond between water molecule and alcohol molecule.

ii. Intra molecular hydrogen bonding

Intra molecular hydrogen bond is formed between hydrogen and electro negative atoms present within the molecule. Example (i) O-nitro phenol

- Hydrogen bond plays an important role in the double helical structure of DNA, and other biological molecules.

<u>**Anomalous properties of Water**</u>

Unique properties of water

1. Density of solid state [ice] is less than that of liquid state [water]

2. Water contracts when heated between $0°$ and $4°C$. This is again unusual because most substance expand when heated in all temperature ranges.

The above mentioned unique properties of water can be explained as due to hydrogen bonding.

1. In ice, the hydrogen bonding between H_2O molecules is more extensive than in liquid water. H_2O molecules are tetrahedrally oriented with respect to one another in solid Ice. Each oxygem atom is surrounded tetrahedrally by four hydrogen atoms, two of these are bonded covalently and the other two by hydrogen bonds. This arrangement gives rise to open cage like structure.

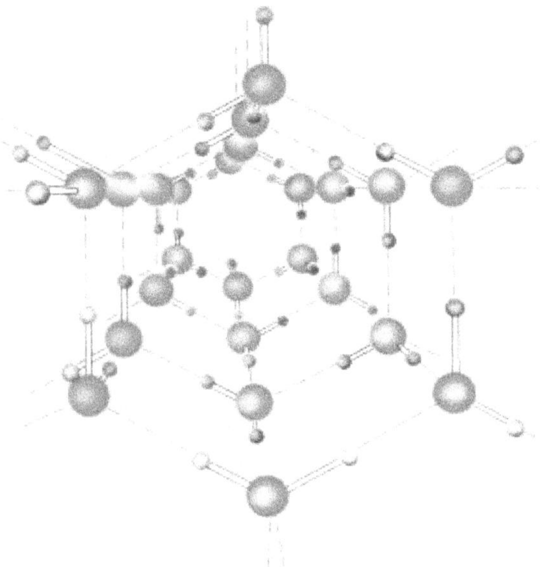

Fig. 27. Hexagonal Structure of Ice

A large number of open spaces or holes are present in this cage like structure of Ice as shown in Fig. 27. Therefore, as ice melts a large number of hydrogen bonds are broken.

The molecule therefore more into the holes or open spaces and come closer to one another than they in solid state. This results in a sharp increase in the density. The density of liquid water is therefore higher than that of ice.

2. As liquid water is heated from $0°$ to $4°C$, hydrogen bonds continue to be broken and the molecules come closer and closer together. This leads to contraction. However, there is some expansion of water also due to rise in temperature as in other liquids. It appears that upto $4°C$, the former effect predominates and hence there is net contraction in volume. Above $4°C$, between the normal expansion effect, due to rise in temperature, predominates and hence volume increases as the temperature rises.

Van der Waals Forces

Vander Waals forces are weak attractive forces which are operative between molecules which are closely packed.These forces exist even in nonpolar diatomic molecule like O_2, N_2, F_2 etc. and monoatomic gas like He.

Vander Walls forces are purely electro static forces and arise from mutual attraction between the nucleus present in one molecule and the electrons present in another molecule.

Vander Waals forces depends on:

a) Number of electrons present in the molecule

b) Molecular mass

c) Physical state

The Vander Waals forces of attraction increases with increase in the size of the molecule. The variation in Vander Waals forces for halogens is in the following order:

$I_2 > Br_2 > Cl_2 > F_2$

Vander Waal force varies in the following order for three different states of matter.

3.11 Theories of Acids and Bases

Acids and Bases

In the early stages of chemistry, acids were distinguished by their sour taste (In Latin **acere** means sour) and their effect on certain pigments such as litmus. Bases were substances which reacted with acids to form salts (Base is alkaline-which is derived from Arabic *alqali)*. Lavoisier believed that the acidic properties of substances were due to the presence of oxygen in them. Davy showed that substances like HCl, H_2S etc. were also acids. Leibig defined acids as the compounds which contained one or more replaceable hydrogen atoms. There were several concepts of acids and bases. The three important theories of acids and bases are:

1. Arrhenius Concept
2. Bronsted and Lowry Concept
3. Lewis Concept

Arrhenius Concept [Water-ion concept]

The Swedish chemist Svante Arrhenius (1884) defined acids and bases in terms of ions given by them. According to Arrhenius, an **acid is a substance which dissociates to give H^+ ions in aqueous solution.**

Example: HCl, HNO_3, H_2SO_4, CH_3COOH etc.

$$HCl \xrightarrow{H_2O} H^+(aq) + Cl^-(aq)$$

$$HNO_3 \xrightarrow{H_2O} H^+(aq) + NO_3^-$$

$$H_2SO_4 \xrightarrow{H_2O} H^+(aq) + HSO_4^-(aq)$$

$$CH_3COOH \xrightarrow{H_2O} CH_3COO^-(aq) + H^+$$

Base is a substance which dissociates to give OH⁻ ions in aqueous solution.

Example: NaOH, KOH, NH_4OH etc.

$$NaOH \xrightarrow{H_2O} Na^+(aq) + OH^-(aq)$$

$$KOH \xrightarrow{H_2O} K^+(aq) + OH^-(aq)$$

$$NH_4OH \rightleftharpoons NH_4^+(aq) + OH^-(aq)$$

Merits

1. This theory explains the common acid - base reactions in aqueous solution. The acid-base neutralization process is actually the combination of H^+ ions and OH^- ions.

$$HCl\ (aq) + NaOH\ (aq) \longrightarrow NaCl\ (aq) + H_2O\ (l)$$

$$H^+(aq) + Cl^-(aq) + Na^+(aq) + OH^-(aq) \longrightarrow Na^+(aq) + Cl^-(aq) + H_2O\ (l)$$

$$H^+(aq) + OH^-(aq) \longrightarrow H_2O\ (l)$$

The enthalpy of neutralization of strong acid by **strong acid by** strong base is 57.3 kJ mol^{-1}. This is actually enthalpy of formation of H_2O by the combination of 1 mole of H^+ ions and 1 mole of OH^- ions. Since strong acids and strong bases dissociate almost completely, 1 mole of acid and base produces 1 mole of H^+ and 1 mole of OH^- ions, which release 57.3 kJ heat on combination.

2. This concept helps in the determination of relative strengths of acids and bases.

Limitations

1. This concept is applicable only for aqueous solutions. The acid-base reactions occurring in non-aqueous solvents cannot be explained on the basis of this concept.
2. This theory fails to explain the basic behaviour of metal oxides.
3. This theory fails to explain the acidic behaviour of certain salts like $AlCl_3$, NH_4Cl etc in water.

BRONSTED - LOWRY CONCEPT [protonic concept]

In 1923, Bronsted-Lowry independently defined acids as proton donars and bases as proton acceptors. H^+ ion is called a proton.

Acids:

According to this concept, acids may be defined as follows:

The molecules, cations or anions which donate [H^+ ions] or [Protons] to any other substance are known as acids.

Example:

$$HCl \longrightarrow H^+ + Cl^-$$
molecular acid

$$[Fe(H_2O)_6]^{3+} \longrightarrow H^+ + [Fe(H_2O)_5(OH)]^{2+}$$
cationic acid

$$HSO_4^- \longrightarrow H^+ + SO_4^{2-}$$

$$HCO_3^- \longrightarrow H^+ + CO_3^{2-}$$

anionic acids

Bases:

The molecular cations or anions which accept protons [H$^+$ ions] from other substances are known as Bases.

$$NH_3 + H^+ \longrightarrow NH_4^+$$

molecular base

$$[Al(H_2O)_5 OH]^{2+} + H^+ \longrightarrow [Al(H_2O)_6]^{3+}$$

cationic   base

$$HSO_4^- + H^+ \longrightarrow H_2SO_4$$

anionic   base

Merits

1. The Bronsted-Lowry theory extends the scope of acid-base systems to cover solvents such as liquid ammonia, acetic acid (glacial), Anhydrous H_2SO_4 and oil hydrogen containing solvents.
2. It should be emphasised that bases accepts proton and there is no need for them to contain OH⁻, the theory qualifies large number of substances as 'base'. E.g., Anions, pyridine, pyrrole, $C_2H_5O^-$ etc.
3. The concept explains the phenomenon of salt hydrolysis.
4. Relative strengths of acids and bases can be determined with the help of this concept.

Limitations

1. This concept cannot explain the non-protonic acid-base reactions.
2. This theory is applicable only for protonic exchange reactions, but fails to explain several acid-base reactions.

$$\text{a.} \quad SO_3 + CaO \longrightarrow CaSO_4$$

$$\text{b.} \quad SO_3 + H_2O \longrightarrow H_2SO_4$$

LEWIS CONCEPT [Electronic Concept]

Lewis developed a definition of acids and bases that did not depend on the presence of protons, nor involve reactions with the solvent. This is the most widely used concept due to its simplicity and wider applications.

Acid

According to Lewis concept, An acid is the substance which accepts electron pair from a base to form a bond [Co-ordinate]

Ex: H$^+$, BF$_3$, Co^{3+}, AlCl$_3$

$$H^+ + {:}NH_3 \longrightarrow H \overset{+}{\underset{\leftarrow}{}} {:}NH_3 \quad \text{or} \quad [H \leftarrow NH_3]^+$$

$$BF_3 + {:}NH_3 \longrightarrow H_3N{:} \longrightarrow BF_3$$

$$Co^{3+} + 6Cl^- \longrightarrow [Co\,Cl_6]^{3-}$$

$$AlCl_3 + Cl^- \longrightarrow AlCl_4^-$$

Base:

According to Lewis concept, Base is the substance which donates an electron pair to other substance to form a bond.

Ex:NH_3, OH^-, Cl^-, RNH_2, CN^-

$$2NH_3 + Ag^+ \longrightarrow [Ag(NH_3)_2]^{2+}$$

$$4NH_3 + Cu^{2+} \longrightarrow [Cu(NH_3)_4]^{2+}$$

$$6CN^- + Fe^{3+} \longrightarrow [Fe(CN)_6]^{3-}$$

Lewis definition is a wider definition and includes those reactions also in which no ions are formed and no transference of hydrogen ion or other ions takes place.

Lewis definition covers all reactions involving H^+ ions, oxide ion (or) solvent interactions as well as the formation of acid-base adducts [complexes] such as

$$R_3N \longrightarrow BF_3 \text{ and all co-ordinate compounds.}$$

Merits
1. This concept is applicable for acid-base reaction which donot involve protons.
2. Basic properties of metal oxides and acidic properties of non-metal oxides can easily be explained on the basis of this concept.
3. This concept includes many gas phase and high temperature non-solvent reactions as neutralization processes.

Limitations
1. This concept has no uniform scale of acid base strength, since the strength of an acid or a base compound is not constant and varies from one solvent to another and also from one reaction to another.
2. Almost all reactions become acid base reactions under this system.

Year	Thinker	Acid	Base	Acid-base reaction
1884	Arrhenius	ionize H^+	ionize OH^-	$H^+ + OH^- = HOH$
1923	Bronsted- Lowry	proton donor	proton acceptor	HA + B = HB + A conjugation
1923	Lewis	electrophile	nucleophile	E + Nu = E:Nu

Conjugate Pairs of Acid and Base

An acid donates a proton and become a base and base Accepts a proton into a base after releasing one proton.

Conjugate acid-base pair; A species formed from an acid by the loss of proton is called the conjugate base of that acid.

Ex: HCl and Cl^- are conjugate acid-base pair

$$HCl \longrightarrow H^+ + Cl^-$$
$$\text{Acid} \qquad \text{conjugate base}$$

Conjugate base-acid pair; A species formed by the gain of a proton by a base is called the conjugate acid of that base.

$$NH_3 + H^+ \longrightarrow NH_4^+$$
$$\text{Base} \qquad \text{Conjugate acid}$$

Ex: NH_3 and H^+ are conjugate acid-base pair

An acid base pair which differs by just a proton is called a conjugate pair or conjugate acid-base pair.

In a conjugate pair, if one of the species is strong, then the other will be weak. A strong acid like HCl has a weak base.

$$HCl \longrightarrow H^+ + Cl^-$$
$$\text{Strong acid} \qquad \text{Weak base}$$

Similarly a weak acid has a strong conjugate base

$$CH_3COOH \longrightarrow CH_3COO^- + H^+$$
$$\text{Weak acid} \qquad \text{Strong base}$$

Consider the following reaction

$$HCl\,(aq) + H_2O\,(l) \rightleftharpoons H_3O^+\,(aq) + Cl^-\,(aq)$$

In this reaction, HCl donates a proton to change into Cl^-. Thus HCl is an acid and Cl^- is its conjugate base. H_2O molecule accepts a H^+ ion and H_3O^+ is its conjugate acid. Since H_3O^+ has a tendency to give a proton and Cl^- has a tendency to accept it, the exchange of proton is reversible process.

$$HCl + H_2O \rightleftharpoons H_3O^+ + Cl^-$$
$$\text{acid}-1 \quad \text{base}-2 \qquad \text{acid}-2 \quad \text{base}-1$$

Other examples are given below:

$$H_2O + NH_3 \rightleftharpoons NH_4^+ + OH^-$$

$$CH_3COOH + H_2O \rightleftharpoons CH_3COO^- + H_3O^+$$

Examples of conjugate pairs

HCl	Cl⁻	H_2S	HS⁻
HNO_3	NO_3^-	H_2CO_3	HCO_3^-
H_2SO_4	HSO_4^-	HCO_3^-	CO_3^{2-}
HSO_4^-	SO_4^{2-}	$[Fe(H_2O)_6]^{3+}$	$[Fe(H_2O)_5OH]^{2+}$
H_3O^+	H_2O	HS⁻	S^{2-}
NH_4^+	NH_3	H_2O	OH⁻

Strengths of Acids and Bases

The strength of an acid is determined by its tendency to lose protons. Higher the tendency of an acid to lose the proton, greater is its strength.

- Acids such as HCl, HNO_3, H_2SO_4 etc possess very high tendency to lose protons. Therefore, they are strong acids.
- Acids such as CH_3COOH, HCN etc., possess little tendency to lose protons and are called weak acids.
- The strength of a base is determined by its tendency to gain protons.
- Bases such as CH_3COO^-, CN^-, OH^-, NH_3 etc are strong bases because they possess very high tendency to gain proton.

Bases such as Cl^-, NO_3^-, SO_4^{2-} etc are weak bases because they have little tendency to accept protons.

The conjugate of a strong acid is always a weak base and the conjugate of a weak acid is always a strong base

$$HCl_{(Strong\ acid)} + H_2O \longrightarrow H_3O^+ + Cl^-_{(Weak\ base)}(aq)$$

$$CH_3COOH_{(Weak\ acid)} + H_2O \rightleftharpoons H_3O^+ + CH_3COO^-_{(Strong\ base)}(aq)$$

Relative Strength of Acids and Bases: Dissociation Constants

I. The strength of an acid may be defined as the concentration of H^+ ions [H_3O^+ ions] in its aqueous solution at a particular temperature.

The dissociation of an acid HA can be written as follows:

$$HA\ (aq) \rightleftharpoons H^+\ (aq) + A^-\ (aq)$$

Applying the law of mass action to this equilibrium

$$K_a = \frac{[H^+][A^-]}{[HA]}$$

K_a is called the dissociation constant or ionisation constant of the acid. For a given acid at a given temperature K_a is a constant. The value of K_a for an acid increases when the temperature increases.

Smaller is the value of dissociation constant, weaker will be the acid

For example, K_a of CH_3COOH is 1.8×10^{-5}

K_a of HF is 6.7×10^{-4}

Since the dissociation constant of HF is greater than that of CH_3COOH, HF is a stronger acid as compared to acetic acid.

II. The strength of a base depends on its tendency to gain protons.

The dissociation of a base BOH can be written as follows:

$$BOH \ (aq) \rightleftharpoons B^+(aq) + OH^-(aq)$$

Applying law of masses action to this equilibrium

$$K_b = \frac{[B^+][OH^-]}{[BOH]}$$

Where K_b is called the dissociation constant or Ionisation constant of base. Higher the value of K_b greater is the strength of the base.

P^{ka} values for weak acids

The P^{ka} value of weak acid may be defined as the negative logarithm (base 10) of the dissociation constant ka of weak acid.

$P^{ka} = -\log_{10}Ka$

For example, k_a of $CH_3COOH = 1.8 \times 10^{-5}$

$$P^{ka} = -\log_{10}ka = -\log_{10}[1.8 \times 10^{-5}]$$

$$= 5 - \log_{10}1.8$$

$$= 5 - 0.2553 = 4.744$$

The smaller the value of P^{ka}, the stronger is the acid.

Formic acid [$P^{ka} = 3.752$] is stronger acid than acetic acid [$P^{ka} = 4.744$]

P^{kb} values for weak bases

P^{kb} value of weak base may be defined as the negative logarithm to the base 10 of the dissociation constant, K_b of the base.

For example,

K_b of $NH_4OH = 1.75 \times 10^{-5}$

$P^{kb} = -\log_{10} K_b = -\log_{10} 1.75 \times 10^{-5}$

$$= 5 - \log_{10} 1.7 \quad = 5 - 0.243 \quad = 4.7575$$

The smaller the value of P^{kb}, the stronger is the base.

Methyl amine $[P^{Kb} = 3.1487]$ is stronger base than NH_4OH $[P^{Kb} = 4.757]$

Note:

1. *Decreasing order of acidic character*

   P^{ka} *values :* $\qquad$ *3.75* $\qquad\qquad$ *4.19* $\qquad\qquad$ *4.74* $\qquad\qquad$ *9.96*

   Acids $\qquad$: $\qquad HCOOH > C_6H_5COOH > CH_3COOH > C_6H_5OH$

2. *Decreasing order of basic character*

   P^{kb} *values :* $\qquad$ *3.22* $\qquad\qquad$ *3.35* $\qquad\qquad$ *4.22* $\qquad\qquad$ *4.76* $\qquad\qquad$ *9.37*

   Bases $\qquad$: $\qquad (CH_3)_2 NH > CH_3NH_2 > (CH_3)_3N > NH_3 > C_6H_5NH_2$

3. $P^{ka} + P^{kb} = 14$

   P^{ka} *of acetic acid = 4.7 .* P^{kb} *of it's conjugate base* CH_3COO^- *is calculated as follows:*

   $P^{kb} = 14 - P^{ka} = 14 - 4.7 = 9.3$

Ionic Product of Water

Water is a weak electrolyte and does not ionize much in the pure state. Water undergoes self ionization as shown below:

Applying the law of mass action to this equilibrium

$\qquad$ or $\quad K[H2O] = [H+] [OH-]$

Since water undergoes ionization to a very small extent, $[H2O]$ may be regarded as constant.

$\qquad K[H2O]$ is a constant. Thus we have

$\qquad$ The constant KW is known as ionic product of water.

$K_W = 10^{-14} \ mol^2/L^2$ at 298 K. Thus in pure water at 298 K,

$\qquad [H+] [OH-] = 10^{-14}$

$\qquad [H+] = [OH-] =$

At 25°C

$\qquad [H+] = 10^{-7} \ mol/dm^3$

$\qquad [OH-] = 10^{-7} \ mol/dm^3$

$\qquad KW = 10^{-7}.10^{-7} = 10^{-14} \ mol^2/(dm^3)^2$

The ionic product of water varies with temperature.

Concentration of H^+ and OH^- ions in aqueous solution of Acids and Bases

When an acid or base is added to pure water, the concentration of H^+ ions and OH^- ions changes but the value of ionic product of water i.e., K_W remains unchanged so long as the temperature remains constant. $\{K_W = [H^+][OH^-] = 10^{-7}.10^{-7} = 10^{-14}\ mol^2/(dm^3)^2\}$

Addition of an acid in water decreases the $[OH^-]$ according to the relation.

$$[OH^-] = \frac{K_W}{[H^+]}$$

Addition of base in water decreases $[H^+]$ according to the relation

$$[H^+] = \frac{K_W}{[OH^-]}$$

Thus H^+ and OH^- ions are always present in an aqueous solution no matter whether it is neutral, acidic or basic. However their relative concentrations are different in different types of solutions. In general,

In neutral solutions; $[H^+] = [OH^-]$

In acidic solutions; $[H^+] > [OH^-]$

In basic solutions; $[H^+] < [OH^-]$

3.12 EXERCISE QUESTIONS

Exercise Questions: Type 1

1. Define a chemical bond.Noble gases do not form compounds. Give reason.

2. Give one compound which is formed due to electrovalent bond/covalent bond/coordinate bond.

3. Draw electron dot structure of $H_2/Cl_2/N_2/O_2/H_2O/NH_3/CH_4$.

4. Give one example of a polar molecule containing chlorine/oxygen/nitrogen.

5. How many π bonds are present in $N_2/O_2/C_2H_4/C_2H_2$?

6. Write the bond angles present in water/methane/ammonia/ethane/ethene/ethyne.

7. Which molecule has zero dipole moment among HF, HCl, NH_3, CH_4?

8. Draw a diagram to show the orbital overlap to form sigma bond and pi bond.

9. What is the shape of $BeCl_2$ molecule?

10. Mention the hybridization involved in methane/ethane/ethene/ethyne/benzene molecule.

11. How are the two carbon-carbon bonds in ethylene different?

12. What is the hybridization present in BF_3?

13. What are the types of bonds present in ammonium ion?

14. Give one example of a compound which exceeds octet structure.

15. Mention two compounds containing which exceeds octet structure.

16. At what temperature has water a maximum density?

17. Why do atoms combine to form molecules?

18. Why are the rare gases monoactomic?

19. What is meant by van der Waal's force?

20. Predict the shape of the molecules of H_2O / NH_3 / BF_3 / SiF_4.

21. Show the polar nature of water molecule.

22. What is meant by dipole moment?

23. Give a molecule with zero dipole moment.

24. Mention hybridization present in diamonds/graphite/methane.

25. How many π-bonds and σ-bonds are present in ethylene?

26. π-bond is weaker than σ-bond. Give reason.

27. What is the shape of BF_3 molecule?

28. CO molecule has dipole moment, but CO_2 does not. Justify.

29. In General, ionic compounds have very high melting points. Give reason.

30. Boiling point of water is abnormally high. Justify.

31. What is polarity?

32. Mention the bonds involved in ammonium chloride molecules.

33. What is meant by inter molecular forces?

34. What are Van der Waal's forces?

35. What is meant by lattice energy?

Exercise Questions: Type 2

36. Explain covalent bond with an example.

37. Explain the formation of NaCl.

38. What are the conditions for the formation of ionic bond/covalent bond?

39. What are the characteristic features of ionic compound/covalent compound?

40. Explain the formation of oxygen molecule.

41. Define coordinate bond. Give an example.

42. Mention the differences between sigma and pi bonds.

43. Explain hybridization with an example.

44. An ionic bond is non-directional where as covalent bond is directional.

45. What is meant by Born-Haber cycle?

46. What are polar and non-polar molecules?

47. Explain shape of BF_3 molecule using VSEPR theory.

48. Explain the polar nature of NH_3 / HF.

49. BF_3 has zero dipole moment. Explain

50. What is the effect of Van der Waal's force on the boiling points of compounds?

51. What is meant by hydration?

52. Explain hydrogen bonding with an example.

Exercise Questions: Type 3

53. Explain ionic bond formation with an example. Mention the factors which favours ionic bond formation.

54. Explain the structures and bondings in methane / ethylene / ethyne / BF_3.

55. Explain the characteristic features of ionic / covalent / coordinate compounds.

56. What is meant by hybridization? Explain the various types of hybridization with examples.

57. What is meant by VSEPR theory? Explain bonding in NH_3 and water using VSEPR theory.

58. How is lattice energy calculated for an ionic crystal using Born Haber cycle?

59. What is meant by dipole moment? Mention any two molecules with zero dipole moment.

60. What are Van der Waal's forces? Explain the effect of molecular mass on Van der Waal's forces.

61. Explain breifly the following acid base theories:

    a. Arrhenius theory

    b. Bronsted Lowry theory

    c. Lewis theory

Chapter 4

TRANSITION ELEMENTS CHEMISTRY (*d*- BLOCK ELEMENTS)

4.1 INTRODUCTION

All the d-block elements are metals and they exhibit several interesting properties. These include precious metals such as silver, gold and platinum and industrially important metals such as iron, copper and titanium. Many transition metals form co-ordination complexes which find application in catalysis, polymerization, medicine etc. Transition metals are used in several industries such as, automobile, aircraft, railways and jewellary etc. In addition to being important in industry, transition metal ions play a significant role in executing number of processes in living organisms. For example, coordination complexes of iron involve in the process of oxygen transport and its storage, molybdenum and iron compounds act as catalysts in nitrogen fixation, zinc is found in more than 150 biomolecules in humans, copper and iron play a crucial role in the respiratory cycle and cobalt is found in essential biomolecules such as vitamin B_{12}.

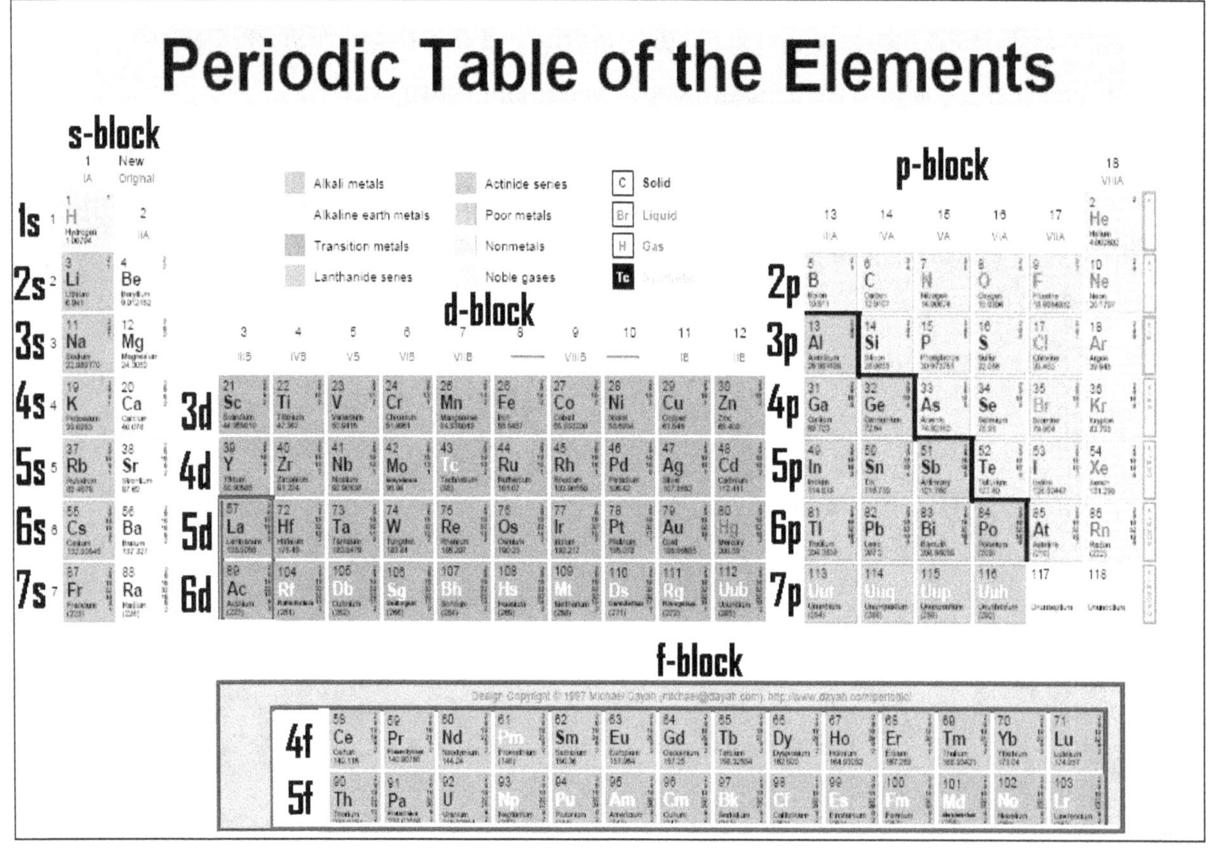

The elements in the periodic table are classified as s-block, p-block, d-block and f-block elements. The s-block and p-block elements are called representative elements or main group elements. The d-block elements are also called transition elements. The f-block elements are called inner-transition elements. Inner transition elements are shown separately, as Lanthanides and Actinides, at the bottom of the periodic table.

d- block elements [Transition elements]

d- block elements are elements with partially filled d-orbitals in their atoms   or The elements in which the last electron enters d-orbitals are known as d-block elements.

d- block elements are present between s-block and p-block elements in periodic table. The d-block elements are called transition elements because they exhibit transitional [intermediate] behaviour between electropositive [Ionic] s-block elements and electronegative [covalent] p-block elements.

The IUPAC definition of transition elements is that it is an element that has an incomplete d Subshell in either the neutral atoms or its ions. Thus the group 12 elements Zn, Cd and Hg are d-block elements but are not transition elements.

Classification: All the d-block elements are classified into four series containing thirty nine elements as given below,

1) First transition series or 3d-series:

3d-series consists of 10 elements from scandium [z = 21] to zinc [z = 30]. All these elements belong to fourth period of the periodic table.

2) Second transition series or 4d-series:

4d-series consists of 10 elements from Yttrium [z = 39] to Cadmium [z = 48]. These elements belong to fifth period of periodic table.

3) Third transition series or 5d-series:

5d-series consists of 10 elements from Lanthanum [z = 57] to Mercury [z = 80]. These elements belong to sixth period of periodic table.

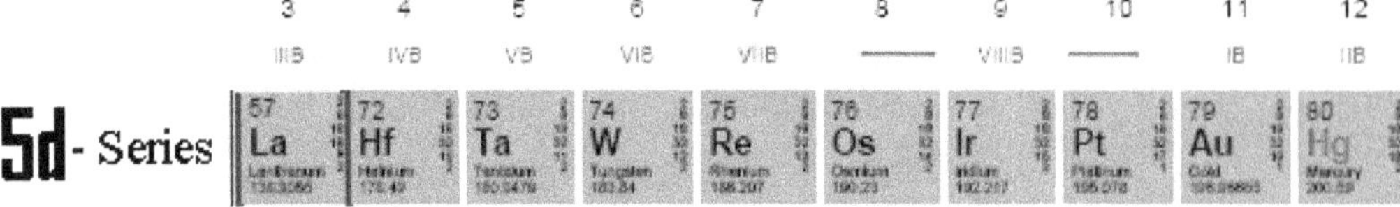

4) Fourth transition series or 6d-series:

6d series starts with Actinium [z = 89]. It is an incomplete seventh period.

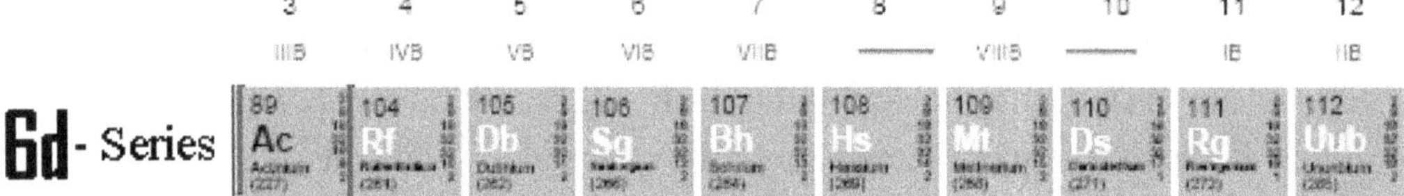

Note:

1. Zinc, Cadmium and Mercury are the last members of 3d, 4d and 5d series respectively. These elements and their ions in +2 state are excluded from transition elements as they do not show properties characteristic of transition elements due to their completely filled d-orbitals.

2. All the transition elements are d-block elements, but all d-block elements are not transition elements.

4.2 GENERAL CHARACTERISTICS OF 3*d* TRANSITION ELEMENTS

The transition element show several interesting physico-chemical properties. The Transition metals show great similarities within a given period as well as within a given vertical group. This is because; the last electrons are added to inner d-orbitals of transition elements. The inner d-electrons cannot participate very easily in bonding. Thus the chemistry of the transition elements is not affected as greatly by the gradual change in the number of electrons as is the chemistry of the main group elements. But also a very important difference, in the properties between the transition elements and main group elements, arises due to the availability of empty d orbitals in d- block elements. Thus the transition elements exhibit the ability to form colored complexes which show excellent magnetic and catalytic behaviours.

The transition metals behave as typical metals, possessing metallic luster and relatively high electrical and thermal conductivities. Silver is the best conductor of heat and electricity. However copper finds very wide electrical applications in domestic as well as industrial sectors due to its position next to silver.

The transition metals do vary significantly in certain properties. For example tungsten has melting point of 3400°C and is used as filaments in bulbs: mercury is a liquid at 25°C. Transition metals such as iron and titanium are hard and strong and make useful structural materials while other metals such as copper, silver and gold are soft. The chemical reactivity of the transition metals also varies significantly. Metals like chromium, nickel and cobalt form oxide which are non porous, less conducting and stoichiometric in nature, protecting the metal from further oxidation. But iron forms oxides, which are porous, highly conducting and unstable in nature, inducing continuous corrosion and total degradation of the metal. On the other hand the noble metals such as gold, silver, platinum and palladium do not readily form oxides.

Trends in Physical Properties

As it is already learnt that the electrons in transition metals enter inner d orbitals, the physical properties of the elements tend to be almost similar. The shapes of the d- orbitals affect the properties of the d-block elements in two ways. First, the electrons repel very weakly in d- orbitals as they are relatively placed apart and the second, the electron density in d-orbitals is low near the nucleus and so they are less effective at shielding other electrons from the positive nuclear charge. The atomic radii of the transition elements in the 3d, 4d and 5d series steadily decrease from left to right across the period. But towards the end of each series there is a slight increase.

	Sc	Ti	V	Cr	Mn	Fe	Co	Ni	Cu	Zn
Atomic radii [A°]	1.44	1.32	1.22	1.17	1.17	1.17	1.16	1.15	1.17	1.25

Atomic radii of transition elements

The atomic radii of 3d-series transition elements (Period4) exhibit the following two trends.

a. <u>The atomic radii of 3d-transition elements decrease with increase in atomic number.</u>

Atomic radii decrease with increase in atomic number due to increase in the nuclear change. In addition, as the atomic number increases, the added electrons enter into 3d-shell and weakly shield the outermost electrons. The shielding effect increases with increase in the number of d-electrons. Thus the effect of the increased nuclear charge due to increase in atomic number is counter balanced by the increased shielding effect of 3d electrons. That is why the atomic radii remain almost constant after midway in the series.

Table 7. Variation of atomic radii of 3d, 4d and 5d series of elements

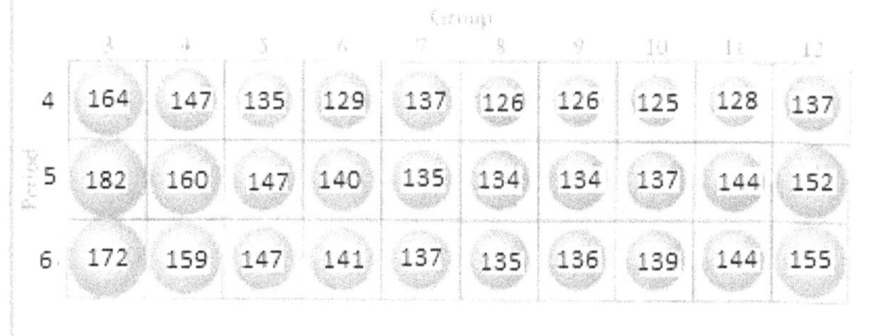

Period	Group									
	3	4	5	6	7	8	9	10	11	12
4	164	147	135	129	137	126	126	125	128	137
5	182	160	147	140	135	134	134	137	144	152
6	172	159	147	141	137	135	136	139	144	155

b. <u>Towards the end of the series, atomic radii increases.</u>

This is due to the greater effect of electron - electron repulsion between the added electrons than the nuclear charge. Thus electron cloud expansion increases atomic radii.

The atomic radii of the second row of d-metals (Period 5) are typically greater than those in the first row (Period 4). The atomic radii in the third row (Period 6) however, are about the same as those in the second row and smaller than expected. This effect is due to the lanthanide contraction, the decrease in

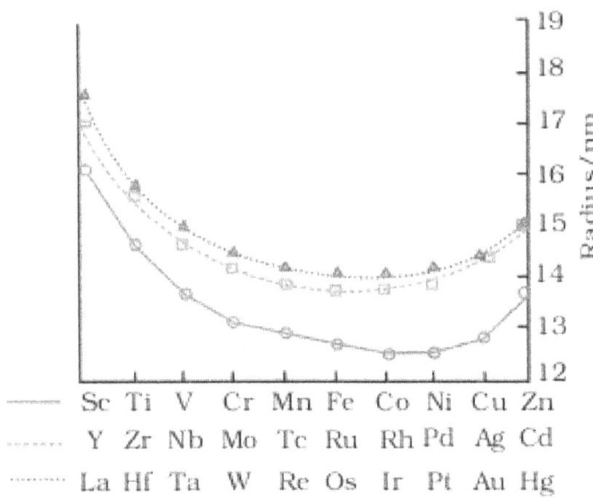

Fig. 28. Trends in atomic radii of transition elements

radius along the first row of the f-block. This decrease is due to the increasing nuclear charge coupled with the poor shielding ability of f- electrons.

The densities of the transition metals are high due to their small sizes and close packed structures. All transition metals have a density greater than 5 g/cm^3. (Exceptions Se-3.0, Y & Ti- 4.5 g/cm^3) In a given transition series, on moving from left to right, the density first increases and then attains the maximum value on reaching group 7.

Element	Group	density /g cm^{-3}	m. p. / °C	b.p. / °C	radius / pm	free atom configuration	ionization energy / kJ mol^{-1}
Sc	3	3.00	1541	2831	164	[Ar] 3d^{1}4s^2	631
Ti	4	4.50	1660	3287	147	[Ar]3d^{2}4s^2	658
V	5	5.96	1890	3380	135	[Ar]3d^{3}4s^2	650
Cr	6	7.20	1857	2670	129	[Ar]3d^{5}4s^1	653
Mn	7	7.20	1244	1962	137	[Ar]3d^{5}4s^2	717
Fe	8	7.86	1535	2750	126	[Ar]3d^{6}4s^2	759
Co	9	8.90	1495	2870	125	[Ar]3d^{7}4s^2	758
Ni	10	8.90	1455	2730	125	[Ar]3d^{8}4s^2	737
Cu	11	8.92	1083	2567	128	[Ar]3d^{10}4s^1	746
Zn	12	7.14	420	907	137	[Ar]3d^{10}4s^2	906

Then it starts decreasing when we move further to group 11 and 12. The densities of the second row are high and third row values are even higher. Down the group, the density increases substantially. The atomic size of the elements in the same group but belonging to second and third transition series remains almost the same but their atomic weight is almost doubled. Two elements with the highest densities are osmium 22.57 g/cm^3 and iridium 22.61 g/cm^3. A football made of osmium or iridium measuring 30 cm in diameter would weigh 320 kg almost one third of a tonne.

The melting points and the molar enthalpies of fusion of the transition metals are both high in comparison to main group elements. This arises from strong metallic bonding in transition metals which occurs due to delocalization of electrons facilitated by the availability of both d and s electrons. All the transition elements have high M.P. and B.P. as compared to s-block elements due to their strong metallic bonding and unpaired d-electrons (Fig. 29). M.P. of transition elements of a given series increases on moving from left to right in a period and attains a maximum value and after that M.P. goes on decreasing towards the end of period. These higher values are due to small atomic radii of transition elements which provides greater inter atomic forces of attraction. However Zn, Cd, Hg have relatively low values of M.P. and B.P. Due to their completely filled d-orbitals

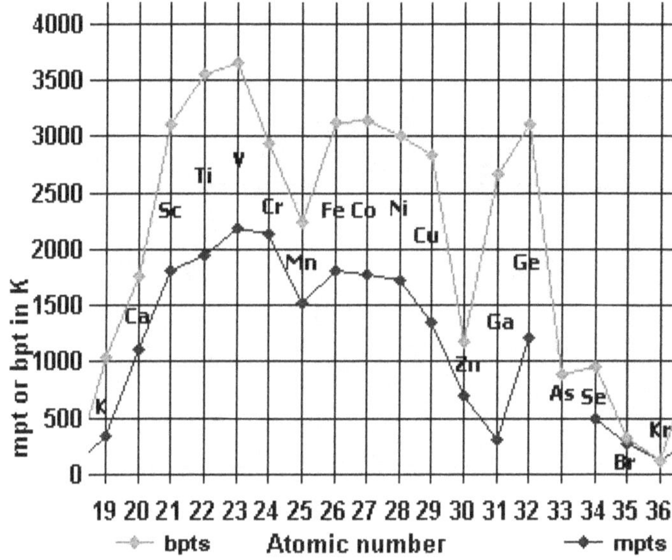

Fig. 29. Melting points and Boiling points for Period 4 elements

Electronic configurations of 3d, 4d, 5d and 6d series of elements

In the first element of 3d series of d-metals, the 3d orbital gets its first electron in scandium. So the electronic configuration of Sc is written as **[Ar] 4s^2 3d^1**. Similarly titanium - **[Ar] 4s^2 3d^2** and vanadium - **[Ar] 4s^2 3d^3** contain 2 and 3 electrons in their 3d orbitals respectively. In the fourth element chromium, the electronic configuration is found to be **[Ar] 3d^5 4s^1** against the predicted one **[Ar] 3d^4 4s^2**.

This is due to the extra stability attained in the first case due to completely half filled configuration of the 3d orbital (As per Hund's rule, electron repulsion is minimum in half filled sub shells). In the similar fashion the electronic configuration of copper is **[Ar] 3d^{10} 4s^1** in the place of **[Ar] 3d^9 4s^2.** The general electronic configuration of d-block elements is ns^2 (n-1) d$^{1-10}$, where n is the outermost orbit. (n-1) is penultimate shell. The electronic configurations of 3d-transition elements are given below:

Element	Sc	Ti	V	Cr	Mn	Fe	Co	Ni	Cu	Zn
Outer electronic configuration	$3d^1 4s^2$	$3d^2 4s^2$	$3d^3 4s^2$	$3d^5 4s^1$	$3d^5 4s^2$	$3d^6 4s^2$	$3d^7 4s^2$	$3d^8 4s^2$	$3d^{10} 4s^1$	$3d^{10} 4s^2$

Chromium and copper attain extra stability by promoting one of the electrons from 4s orbital to 3d orbital. Thus chromium and copper possess half-filled orbital and completely filled d-orbital respectively and attain extra stable state. The number of unpaired electrons in the atoms of transition elements can be obtained by placing the electrons in the orbital diagram as per Hund's rule. For example, the number of unpaired electrons in manganese is 5,

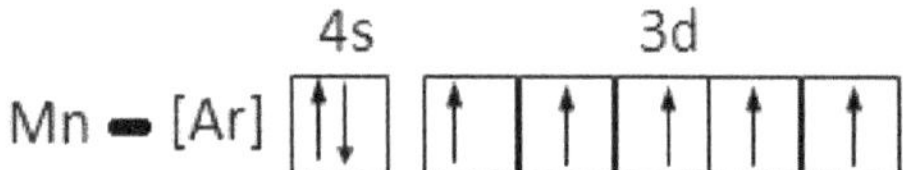

Table 8. Valence electronic configurations of d- block elements

	3	4	5	6	7	8	9	10	11	12
	IIIB	IVB	VB	VIB	VIIB	——	VIIIB	——	IB	IIB
3d	21 Sc $3d^1 4s^2$	22 Ti $3d^2 4s^2$	23 V $3d^3 4s^2$	24 Cr $3d^5 4s^1$	25 Mn $3d^5 4s^2$	26 Fe $3d^6 4s^2$	27 Co $3d^7 4s^2$	28 Ni $3d^8 4s^2$	29 Cu $3d^{10} 4s^1$	30 Zn $3d^{10} 4s^2$
4d	39 Y $4d^1 5s^2$	40 Zr $4d^2 5s^2$	41 Nb $4d^4 5s^1$	42 Mo $4d^5 5s^1$	43 Tc $4d^5 5s^2$	44 Ru $4d^7 5s^1$	45 Rh $4d^8 5s^1$	46 Pd $4d^{10} 5s^0$	47 Ag $4d^{10} 5s^1$	48 Cd $4d^{10} 5s^2$
5d	57 La $5d^1 6s^2$	72 Hf $5d^2 6s^2$	73 Ta $5d^3 6s^2$	74 W $5d^4 6s^2$	75 Re $5d^5 6s^2$	76 Os $5d^6 6s^2$	77 Ir $5d^7 6s^2$	78 Pt $5d^9 6s^1$	79 Au $5d^{10} 6s^1$	80 Hg $5d^{10} 6s^2$
6d	89 Ac $6d^1 7s^2$	104 Rf $6d^2 7s^2$	105 Db $6d^3 7s^2$	106 Sg $6d^4 7s^2$	107 Bh $6d^5 7s^2$	108 Hs $6d^6 7s^2$	109 Mt $6d^7 7s^2$	110 Ds $6d^9 7s^1$	111 Rg $6d^{10} 7s^1$	112 Uub $6d^{10} 7s^2$

The electronic configurations of *4d* and *5d* transition series are given in Table 8. The electronic configuration of first raw transition elements is regular with only exception of Cr ($3d^5 4s^1$) and Cu ($3d^{10} 4s^1$) electronic configuration. However, for second and third row transition elements the electronic configuration is not regular.

There are evidently some pronounced irregularities in the configuration of these elements. Like Cr and Cu the electronic configuration of Mo ($4d^5 5s^1$), Ag ($4d^{10} 5s^1$) Re ($5d^5 6s^2$) and Au ($5d^{10} 6s^1$) can be easily understood on the basis of higher stability of exactly half filled and fully filled *d* orbital. However, this concept cannot explain the anomalous configuration of Nb, Ru, Rh, Pd, W, Pt. As a matter of fact, no simple explanation for such anomalies can be offered. There are two factors which play a significant role in determining these configurations.

(1) Electron-Nuclear attraction

(2) Electron-Electron repulsion.

Variable oxidation states

One of the most striking features of the transition elements is that the elements usually exist in several different oxidation states, leading to very rich and fascinating to chemistry. Transition elements have the following general electronic configuration ns^{1-2} $(n-1)d^{1-10}$. S -orbital electrons are easily lost to give +2 oxidation state. Higher oxidation states are expected for transition elements due to the loss of electrons from d-orbitals also. The energy difference between the $(n-1)d$ and ns orbitals is very small.(i.e. between the two orbital 3d and 4s,4d and 5s, 5d and 6s). Consequently the ionization energy for the removal of electron from these orbitals is quite similar. As a result transition elements form variable oxidation states.

Table 9. Oxidation states of 3*d* series of elements

Element	Sc	Ti	V	Cr	Mn	Fe	Co	Ni	Cu	Zn
Outer electronic configuration	$3d^1 4s^2$	$3d^2 4s^2$	$3d^3 4s^2$	$3d^5 4s^1$	$3d^5 4s^2$	$3d^6 4s^2$	$3d^7 4s^2$	$3d^8 4s^2$	$3d^{10} 4s^1$	$3d^{10} 4s^2$
Oxidation states	**+3**	**+2**	**+2**	**+2**	**+2**	**+2**	**+2**	**+2**	**+1**	**+2**
		+3	**+3**	**+3**	+3	**+3**	**+3**	+3	**+2**	
		+4	+4	+4	**+4**	+6				
			+5	+5	+5					
				+6	+6					
					+7					

The first transition series (3d) metals oxidation states are given in the table below. The most common oxidation states for the metals are bolded and underlined. The first element loses all the outer electrons to attain +3 oxidation state which is more stable. In aqueous solution titanium form $Ti(H_2O)_6^{3+}$, which easily get oxidized to Ti^{4+}. Titanium is not stable in its +2 state in aqueous solution but exhibits stability in the solid TiO form.

Vanadium exhibits principal oxidation state, +5 in V_2O_5 compound. The oxidation states from +2 to +5 all exist in aqueous solution. The higher oxidation states, +4 and +5 do not exist as hydrated ions as V^{n+} (aq) because the highly charged ion causes the attached water molecules to be acidic. This is followed by the loss of H+ ions to give oxy cations VO_2^+ and VO^{2+}. The most common oxidation states of chromium are +2, +3 and +6. Manganese exists in all oxidation states from +2 to +7 states, although +2, +4 and +7 are the most common ones. The most usual oxidation states of iron and cobalt are +2 and +3. Nickel and zinc are most stable in their +2 oxidation state. Copper is available in two different oxidation states, cupric (+2) and cuprous (+1) forms.

Important conclusions about the variable oxidation states of transition metals are:

1) The increase in oxidation states is observed on moving from Sc (+3) to Mn (+7) but the decrease in oxidation is seen from Mn (+7) to Cu (+1) due to the pairing of electrons
2) Oxidation state +3 is common to the left of the 3d series and +2 is common for metals from the middle to the right of the block.
3) The bonds formed by transition metals in +2 and +3 oxidation states are mostly ionic. On the other hand, the bonds formed in the higher oxidation states are essentially covalent. Thus the bonds in +2 and +3 oxidation states are generally formed by the loss of two or three electrons respectively, while the bonds in higher oxidation states are formed by sharing of d-electrons. For example MnO_4- (Mn in +7) state all the bonds are covalent.
4) The highest oxidation states of transition metals are found in their fluorides and oxides. Re and Os shows highest oxidation state of +8 in ReO_4 and OsO_4. Down the group the stability of high oxidation states increases.
5) After scandium d-orbitals become more stable than s-orbital. Therefore +2 state is unstable for scandium.
6) Transition metals also form compounds in low oxidation state such as -1 and 0.

The group oxidation number of an element is its group number. This group oxidation number is achieved by the elements that lie towards the left of the d block but not by the elements on the right. For example, scandium, yttrium and lanthanum display +3 oxidation state, which corresponds to the loss of all the electrons from their outer 3d and 4s orbitals. The group oxidation state is not achieved in Groups 8-12 of the 3d series (Fe, Co, Ni, Cu and Zn). The most stable oxidation states that are most stable under acid conditions are Ti^{3+}, V^{3+}, Cr^{3+}, Mn^{2+}, Fe^{2+}, Co^{2+} and Ni^{2+}.

The elements of second and third row do not show identical pyramid of oxidation states as the first row. In iron family Os and Ru show oxidation states upto (VIII) for example OsO_4 and RuO_4. The elements second and third rows they show maximum oxidation state for example MO_4 (M = Os or Ru) with oxygen and Re (+VII) in ReF_7 with fluorine.

Table 10. Oxidation states of 4*d* and 5*d* series of elements

Element	Y	Zr	Nb	Mo	Tc	Ru	Rh	Pd	Ag	Cd
Oxidation states	+3	+4	+3	+2	+2	+2	+1	+2	+1	+2
			+4	+3	+3	+3	+3	+4		
			+5	+4	+4	+4				
				+5	+5	+5				
				+6	+6	+6				
					+7	+7				
						+8				

Element	La	Hf	Ta	W	Re	Os	Ir	Pt	Au	Hg
Oxidation states	+3	+4	+3	+2	+2	+3	+1	+2	+1	+1
			+4	+3	+3	+4	+3	+4	+3	+2
			+5	+4	+4	+5				
				+5	+5	+6				
				+6	+6	+7				
					+7	+8				

Colour of the transition metal ions

Most of the compounds of transition metals are coloured in the solid state and/or in the solution phase. This is in contrast to the compounds of s- and p-block elements which are usually white. The colour of transition metal ions is due to electronic transitions with in splitted d-orbitals. These transitions are called d-d transitions.

The colour of the compounds of transition metals may be attributed to the presence of (n-1) d-subshell. In the case of compounds of transition metals, energies of five d-orbitals in the same sub-shell do not remain equal. Under the influence of approaching ions towards the central metal ion, the d-orbitals of the central metal split into different energy levels. This phenomenon is called *crystal field splitting*.

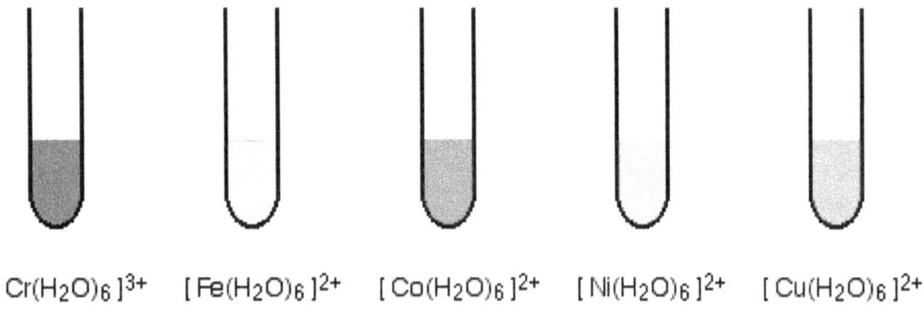

Fig. 30. Colour of transition metal complexes

For example, when six ions or molecules approach the metal ion (called octahedral field) , the five d-orbitals split into two sets : one set (e_g) consisting of two d-orbitals ($d_{x^2-y^2}$, d_{z^2}) of higher energy and the other set (t_{2g}) consisting of three d-orbitals (d_{xy}, d_{yz}, d_{xz}) of lower energy. The relative energies during splitting depend upon the geometry of ligands around the central metal ion and nature of the metal ion.

In the simplest case when the light is absorbed by a complex ion, an electron is absorbed by a complex ion, an electron in one of the lower energy t_{2g} orbital is excited to higher energy e_g orbitals. The energy corresponding to the frequency of absorbed light is equal to Δ.

4.3 CHEMISTRY OF FIRST ROW TRANSITION ELEMENTS AND THEIR COMPOUNDS

Occurrence, Extraction, Properties and Applications

4.3.1. Scandium: Scandium is a chemical element with symbol Sc and atomic number 21. A silvery-white metallic transition metal.

- **Occurrence of Scandium:** Scandium (Fig. 31) is widely distributed in Earth's crust. It is extracted from the ore **thortveitite**, $Sc_2Si_2O_7$. Thortveitite can contain up to 45% of scandium in the form of scandium (III) oxide. Scandium is only the 50th most common element on earth (35th most abundant in the crust), but it is the 23rd most common element in the Sun.

Fig. 31. Scandium metal

- **Extraction of Scandium:** Scandium is prepared by the electrolysis of $ScCl_3$. The production of metallic scandium is in the order of 10 kg per year. The oxide is converted to scandium fluoride and reduced with metallic calcium.

- **Properties of Scandium:** Scandium, despite being a rare element, its physical and chemical properties have been well characterized. Scandium exists in compounds mainly in the +3 oxidation state- for example, in $ScCl_3$, Sc_2O_3 and $Sc_2(SO_4)_3$. Its chemistry is based mostly on the

Sc^{3+} ion. In many ways, scandium resembles main group elements rather than transition elements. For example, scandium reacts with either acidic or basic solutions and even with water to liberate H_2 (g). Another similarity to aluminium is that scandium hydroxide is amphoteric: $Sc(OH)_3$ reacts with acidic solutions to produce Sc^{3+} and with alkaline solutions to form $[Sc(OH)_6]^{3-}$.

The properties of scandium compounds are intermediate between those of aluminium and yttrium. A diagonal relationship exists between the behavior of magnesium and scandium, just as there is between beryllium and aluminium.

Scandium is a soft metal with a silvery appearance. It develops a slightly yellowish or pinkish cast when oxidized by air. It is susceptible to weathering and dissolves slowly in most dilute acids. It does not react with a 1:1 mixture of nitric acid (HNO_3) and 48% hydrofluoric acid (HF), possibly due to the formation of an impermeable passive layer. Scandium turnings ignite in air with a brilliant yellow flame to form scandium (III) oxide.

- **Applications of Scandium:**
    1. The main application of scandium by weight is in aluminium-scandium alloys for minor aerospace industry components. These alloys contain between 0.1% and 0.5% of scandium. They were used in the Russian military aircraft, specifically the MiG-21 and MiG-29.
    2. Scandium-aluminium alloys find use in the manufacture of baseball bats, and bicycle frames and components.
    3. Scandium is used to make high intensity discharge lamps.
    4. The radioactive isotope ^{46}Sc is used in oil refineries as a tracing agent.
    5. Scandium trifluoromethanesulfonate, commonly called scandium triflate, is a chemical compound with formula $Sc(SO_3CF_3)_3$. It is a catalytic Lewis acid used in organic chemistry.

- **Compounds of Scandium:**

The chemistry is almost completely dominated by the trivalent ion, Sc^{3+}. In terms of chemical properties, scandium ions are more similar to those of yttrium than to those of aluminium. In part for this similarity, scandium is often classified as a lanthanide-like element.

Oxides and hydroxides

The oxide Sc_2O_3 and the hydroxide $Sc(OH)_3$ are amphoteric:

$$Sc(OH)_3 + 3\ OH^- \rightarrow Sc(OH)_6^{3-}$$
$$Sc(OH)_3 + 3\ H^+ + 3\ H_2O \rightarrow [Sc(H_2O)_6]^{3+}$$

The α- and γ- forms of scandium oxide hydroxide (ScO(OH)), are isostructural with their aluminium oxide hydroxide counterparts. Solutions of Sc^{3+} in water are acidic because of hydrolysis.

Halides and pseudohalides

The halides ScX_3 (X = Cl, Br, I) are very soluble in water, but ScF_3 is insoluble. In all four halides the scandium is 6-coordinated. The halides are Lewis acids; for example, ScF_3 dissolves a solution containing excess fluoride to form $[ScF_6]^{3-}$. The coordination number 6 is typical of Sc (III). In the larger

Y^{3+} and La^{3+} ions, coordination numbers of 8 and 9 are common. Scandium (III) triflate is sometimes used as a Lewis acid catalyst in organic chemistry.

Organic derivatives
Scandium forms a series of organometallic compounds with cyclopentadienyl ligands (Cp), similar to the behavior of the lanthanides. One example is the chlorine-bridged dimer, $[ScCp_2Cl]_2$ and related derivatives of pentamethylcyclopentadienyl ligands.

4.3.2. Titanium: Titanium is a transition element with atomic number 22 (Fig. 32). It is a lustrous transition metal with a silver color, low density and high strength. It is highly resistant to corrosion in sea water, aqua regia and chlorine. Titanium was discovered in Cornwall, Great Britain, by William Gregor in 1791 and named by Martin Heinrich Klaproth for the Titans of Greek mythology.

Fig. 32. Titanium metal

- **Occurrence of titanium**

Titanium is always bonded to other elements in nature. It is the ninth-most abundant element in the Earth's crust (0.63% by mass) and the seventh-most abundant metal. It is present in most igneous rocks and in sediments derived from them. This element occurs mainly as oxides, **TiO_2 (rutile) and Perovskite (CaTiO3)**. In these, the titanium occupies octahedral holes.

- **Extraction of titanium:**

Titanium metal is produced commercially by the **Kroll process**, a complex and expensive batch process. In the Kroll process, the oxide is first converted to chloride through carbochlorination, whereby chlorine gas is passed over red-hot rutile or ilmenite in the presence of carbon to make $TiCl_4$. This is condensed and purified by fractional distillation and then reduced with 800 °C molten magnesium in an argon atmosphere.

Most titanium is produced in a two steps process from TiO_2. The first step to form metallic titanium is to convert TiO_2 into $TiCl_4$. TiO2 is heated with carbon and chlorine at about 800°C to form gaseous $TiCl_4$.

$TiO_2(s) + 2\ Cl_2(g) \rightarrow TiCl_4(l) + O_2(g)$
Then the TiCl4 is reduced with magnesium in an argon atmosphere at about1000 °C to produce Ti(s). **2 $Mg(l) + TiCl_4(g) \rightarrow 2\ MgCl_2(l) + Ti(s)$**

- **Properties of titanium:**

A metallic element, titanium is recognized for its high strength-to-weight ratio. It is a strong metal with low density that is quite ductile (especially in an oxygen-free environment), lustrous, and metallic-white in color. The relatively high melting point (1,667 °C) makes it useful as a refractory metal. It is paramagnetic and has fairly low electrical and thermal conductivity. Titanium is fairly hard, non-magnetic and a poor conductor of heat and electricity.

Like aluminium and magnesium metal surfaces, titanium metal and its alloys oxidize immediately upon exposure to air.

- Titanium readily reacts with oxygen at 1,200 °C (2,190 °F) in air, and at 610 °C (1,130 °F) in pure oxygen, forming titanium dioxide.

  $Ti(s) + O_2(g) \rightarrow TiO_2$ (s)

  Titanium is one of the few elements that burn in pure nitrogen gas, reacting at 800 °C (1,470 °F) to form titanium nitride, which causes embrittlement.

  $2Ti(s) + N_2(g) \rightarrow 2TiN$ (s)

Titanium is, however, slow to react with water and air, as it forms a passive and oxide coating that protects the bulk metal from further oxidation. Because of its high reactivity toward oxygen, nitrogen and many other gases, titanium filaments are applied in titanium sublimation pumps as scavengers for these gases. Such pumps are inexpensive and reliable devices for producing extremely low pressures in ultra-high vacuum systems.

- **Applications of Titanium:** Three desirable properties make titanium a highly useful metal:
a. Low density- 4.5 gm / dm^3
b. High structural strength even at high temperature
c. Corrosion resistance

1. Titanium is used in steel as an alloying element (ferro-titanium) to reduce grain size and as a deoxidizer, and in stainless steel to reduce carbon content.

2. Titanium is often alloyed with aluminium (to refine grain size), vanadium, copper (to harden), iron, manganese, molybdenum, and with other metals. Applications for titanium mill products (sheet, plate, bar, wire, forgings, castings) can be found in industrial, aerospace, recreational, and emerging markets.

3. Powdered titanium is used in pyrotechnics as a source of bright-burning particles.

4. Titanium filaments are applied in titanium sublimation pumps as scavengers for oxygen and nitrogen gases. Such pumps are inexpensive and reliable devices for producing extremely low pressures in ultra-high vacuum systems.

- **Compounds of Titanium:**

Oxides and alkoxides of titanium

1. **Titanium dioxide: TiO_2 (Titania)**

The most important oxide is TiO_2, which exists in three important polymorphs, anatase, brookite, and rutile. All of these are white diamagnetic solids, although mineral samples can appear dark (see rutile). They adopt polymeric structures in which Ti is surrounded by six oxide ligands that link to other Ti centers. Titania exists in a number of crystalline forms the most important of which are anatase and rutile. These ores are the principal raw materials used in the manufacture of titanium dioxide pigment. The first step is to purify the ore, and is basically a refinement step. Either the sulphate process, which uses sulphuric acid as an extraction agent or the chloride process, which uses chlorine, may achieve this. After purification the powders may be treated (coated) to enhance their performance as pigments.

Applications

Applications for sintered titania are limited by its relatively poor mechanical properties. It does however find a number of electrical uses in sensors and electrocatalysis. By far its most widely used application is as a pigment, where it is used in powder form, exploiting its optical properties.

a. Pigments

The most important function of titanium dioxide however is in powder form as a pigment for providing whiteness and opacity to such products such as paints and coatings (including glazes and enamels), plastics, paper, inks, fibres and food and cosmetics. Titanium dioxide is by far the most widely used white pigment. Titania is very white and has a very high refractive index – surpassed only by diamond. The refractive index determines the opacity that the material confers to the matrix in which the pigment is housed. Hence, with its high refractive index, relatively low levels of titania pigment are required to achieve a white opaque coating.

The high refractive index and bright white colour of titanium dioxide make it an effective opacifier for pigments. The material is used as an opacifier in glass and porcelain enamels, cosmetics, sunscreens, paper, and paints. One of the major advantages of the material for exposed applications is its resistance to discoloration under UV light.

b. Photocatalysis

Titania acts as a photosensitiser for photovoltaic cells, and when used as an electrode coating in photoelectrolysis cells can enhance the efficiency of electrolytic splitting of water into hydrogen and oxygen.

c. Oxygen Sensors

Even in mildly reducing atmospheres titania tends to lose oxygen and become sub stoichiometric. In this form the material becomes a semiconductor and the electrical resistivity of the material can be correlated to the oxygen content of the atmosphere to which it is exposed. Hence titania can be used to sense the amount of oxygen (or reducing species) present in an atmosphere.

d. Antimicrobial Coatings

When surfaces are coated with titanium dioxide, they become resistant to dirt and bacteria. The photocatalytic activity of titania results in thin coatings of the material exhibiting self cleaning and disinfecting properties under exposure to UV radiation. These properties make the material a candidate for applications such as medical devices, food preparation surfaces, air conditioning filters, and sanitaryware surfaces.

2. Barium Titanate: $BaTiO_3$ (Titanate)

Titanates usually refer to compounds made from titanium dioxide, as represented barium titanate ($BaTiO_3$), which is produced synthetically. With a perovskite structure, this material exhibits piezoelectric properties and is used as a transducer in the interconversion of sound and electricity. Many minerals are titanates, e.g. illmenite ($FeTiO_3$). Star sapphires and rubies get their asterism (star-forming shine) from the presence of titanium dioxide impurities.

3. Titanium ethoxide: $Ti_4(OEt)_{16}$

The alkoxides of titanium (IV) are useful compounds that convert to the dioxide. They are volatile, colourless compounds that are sensitive to water. Titanium ethoxide, with the molecular formula $Ti_4(Oet)_{16}$, reacts with water to deposit solid TiO_2 via the sol-gel process. Titanium isopropoxide is used in the synthesis of chiral organic compounds via the Sharpless epoxidation.

Sulphide of titanium

4. Titanium sulphide: TiS_2

Titanium forms a variety of sulfides, but only TiS_2 has attracted significant interest. It adopts a layered structure and was used as a cathode in early generation of lithium batteries. Since Ti (IV) is a "hard cation", the sulfides of titanium are unstable in the presence of moisture, tending to hydrolyze to the oxide and hydrogen sulfide.

Boride of titanium

5. Titanium boride: TiB_2

TiB_2 is the most stable of several titanium-boron compounds. The material does not occur in nature but may be synthesised by carbothermal reduction of TiO_2 and B_2O_3.

As with other largely covalent bonded materials, TiB_2 is resistant to sintering and is usually densified by hot pressing or hot isostatic pressing. Pressureless sintering of TiB_2 can achieve high densities but liquid forming sintering aids such as iron, chromium and carbon, are required.

TiB_2 is resistant to oxidation in air up to 1000°C. It is also resistant to HCl and HF but reacts with H_2SO_4 and HNO_3. It is readily attacked by alkalis. Hot pressing of TiB_2 (with small additions of metallic or carbide sintering aids) is carried out at 1800 - 1900°C and achieves close to theoretical density. Pressureless sintering requires higher levels of sintering aids and sintering temperatures in excess of 2000°C.

Applications

Due to its high hardness, extreme melting point and chemical inertness, TiB_2 is a candidate for a number of applications.

a. Ballistic Armour

The combination of high hardness and moderate strength make it attractive for ballistic armour, but its relatively high density and difficulty in forming shaped components make it less attractive for this purpose than some other ceramics.

b. Aluminium Smelting

The chemical inertness and good electrical conductivity of TiB_2 have led to its use as cathodes in Hall-Heroult cells for primary aluminium smelting. It also finds use as crucibles for handling molten metals and as metal evaporation boats.

c. Other Applications

High hardness, moderate strength and good wear resistance make titanium diboride a candidate for use in seals, wear parts and, in composites with other materials and cutting tools.

In combination with other primarily oxide ceramics, TiB_2 is used to constitute composite materials in which the presence of the material serves to increase strength and fracture toughness of the matrix.

Nitride of titanium

6. Titanium nitride: TiN

Titanium nitride (TiN), having a hardness equivalent to sapphire and carborundum (9.0 on the Mohs Scale), is often used to coat cutting tools, such as drill bits. It also finds use as a gold-colored decorative finish and as a barrier metal in semiconductor fabrication. Applied as a thin coating, TiN is used to harden and protect cutting and sliding surfaces, for decorative purposes (due to its gold appearance), and as a non-toxic exterior for medical implants. In most applications a coating of less than 5 micrometres (0.00020 in) is applied. A thin film of titanium nitride was chilled to near absolute zero converting it into the first known superinsulator, with resistance suddenly increased by a factor of 100,000.

Carbide of titanium

7. Titanium carbide: TiC

Titanium carbide, TiC, is an extremely hard (Mohs 9-9.5) refractory ceramic material, similar to tungsten carbide. It is commercially used in tool bits. It has the appearance of black powder with NaCl-type face centered cubic crystal structure. It is mainly used in preparation of cermets, which are frequently used to machine steel materials at high cutting speed. The resistance to wear, corrosion, and oxidation of a tungsten carbide-cobalt material can be increased by adding 6-30% of titanium carbide to tungsten carbide.

Halide of titanium

7. Titanium tetrachloride: $TiCl_4$

Titanium tetrachloride (titanium (IV) chloride, $TiCl_4$) is a colorless volatile liquid (commercial samples are yellowish) that in air hydrolyzes with spectacular emission of white clouds. Via the Kroll process, $TiCl_4$ is produced in the conversion of titanium ores to titanium dioxide, e.g., for use in white paint. It is widely used in organic chemistry as a Lewis acid, for example in the Mukaiyama aldol condensation. In the van Arkel process, titanium tetraiodide (TiI_4) is generated in the production of high purity titanium metal.

4.3.3. Vanadium: Vanadium is a chemical element with the symbol **V** and atomic number 23. It is a hard, silvery gray, ductile and malleable transition metal. The element is found only in chemically combined form in nature, but once isolated artificially, the formation of an oxide layer stabilizes the free metal somewhat against further oxidation.

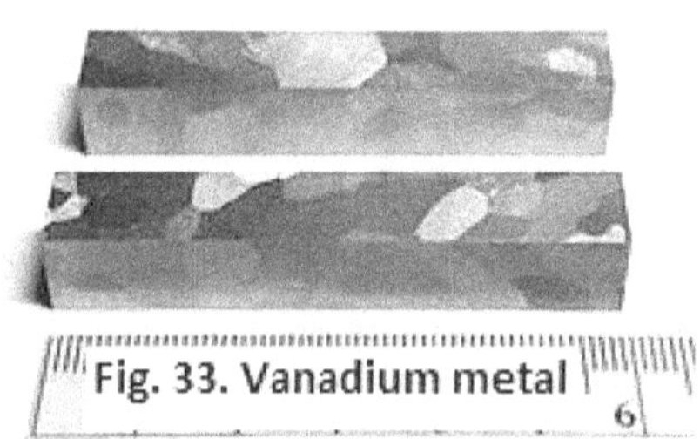

Fig. 33. Vanadium metal

- **Occurrence of vanadium**

Fig. 34. Vanadium bearing Magnetite

Metallic vanadium is not found in nature, but is known to exist in about 65 different minerals. Economically significant examples include **patronite** (VS_4), **vanadinite** ($Pb_5(VO_4)_3Cl$), and **carnotite** ($K_2(UO_2)_2(VO_4)_2 \cdot 3H_2O$). Much of the world's vanadium production is sourced from vanadium-bearing magnetite (Fig.34) found in ultramafic gabbro bodies.

- **Extraction of vanadium**

Vanadium-bearing magnetite iron ore is the main source for the production of vanadium. Vanadium metal is obtained via a multistep process that begins with the roasting of crushed ore with NaCl or Na_2CO_3 at about 850 °C to give sodium metavanadate ($NaVO_3$). An aqueous extract of this solid is acidified to give "red cake", a polyvanadate salt, which is reduced with calcium metal. As an alternative for small-scale production, vanadium pentoxide is reduced with hydrogen or magnesium. Many other methods are also in use, in all of which vanadium is produced as a byproduct of other processes. Purification of vanadium is possible by the crystal bar process developed by Anton Eduard van Arkel and Jan Hendrik de Boer in 1925. It involves the formation of the metal iodide, in this example vanadium (III) iodide, and the subsequent decomposition to yield pure metal.

$2\,V + 3\,I_2 \rightleftharpoons 2\,VI_3$.

- **<u>Properties of vanadium</u>**

Vanadium is a hard, ductile, silver-gray metal. Some sources describe vanadium as "soft", perhaps because it is ductile, malleable and not brittle. Vanadium is harder than most metals and steels. It has good resistance to corrosion and it is stable against alkalis, sulfuric and hydrochloric acids. It is oxidized in air at about 933 K (660 °C, 1220 °F), although an oxide layer forms even at room temperature.

- **<u>Applications of vanadium</u>**

  - Vanadium foil is used in cladding titanium to steel
  - Vanadium-gallium tape is used in superconducting magnets
  - Vanadium pentoxide V_2O_5, is used as a catalyst in manufacturing sulfuric acid by the contact process and as an oxidizer in maleic anhydride production
  - vanadium dioxide VO_2, is used in the production of glass coatings, which blocks infrared radiation (and not visible light) at a specific temperature
  - Vanadate can be used for protecting steel against rust and corrosion by electrochemical conversion coating.
  - Lithium vanadium oxide has been proposed for use as a high energy density anode for lithium ion batteries, at 745 Wh/L when paired with a lithium cobalt oxide cathode
  - Vanadium plays a very limited role in biology. A vanadium nitrogenase is used by some nitrogen-fixing micro-organisms, such as *Azotobacter*. In this role vanadium replaces more common molybdenum or iron, and gives the nitrogenase slightly different properties

- **<u>Compounds of vanadium</u>**

<u>**Oxide of vanadium**</u>

Vanadium Pentoxide

The most commercially important compound is vanadium pentoxide. It has complicated arrangement of trigonal bipyramids. It dissolves in acid to produce the pale yellow VO^{2+} ion. In alkali it forms VO_4^{3-}. It is used as a catalyst for the production of sulfuric acid. This compound oxidizes sulfur dioxide (SO_2) to

the trioxide (SO_3). In this redox reaction, sulfur is oxidized from +4 to +6, and vanadium is reduced from +5 to +4:

$$V_2O_5 + SO_2 \rightarrow 2\,VO_2 + SO_3$$

The catalyst is regenerated by oxidation with air:

$$2\,VO_2 + O_2 \rightarrow V_2O_5$$

Similar oxidations are used in the production of maleic anhydride, phthalic anhydride, and several other bulk organic compounds.

V_2O_3 is highly basic (V (III) ions formed) and is conductive above 170K but insulating below this, VO can also be formed (grey).Thus, wide range of oxides possible.

Oxyanions of vanadium

 In aqueous solution, vanadium (V) forms an extensive family of oxyanions. The tetrahedral orthovanadate ion, VO_4^{3-}, is the principal species present at pH 12-14. Analogies exist between orthovanadate and orthophosphate owing to the similarity in size and charge of phosphorus (V) and vanadium (V). Orthovanadate VO_4^{3-} is used in protein crystallography to study the biochemistry of phosphate. The tetrathiovanadate $[VS_4]^{3-}$ is analogous to the orthovanadate ion.

At lower pH's, the monomer $[HVO_4]^{2-}$ and dimer $[V_2O_7]^-$ are formed, with the monomer predominant at vanadium concentration of less than ca. $10^{-2}M$ (pV > 2; pV is equal to minus the logarithm of the total vanadium concentration/M). The formation of the divanadate ion is analogous to the formation of the dichromate ion. As the pH is reduced, further protonation and condensation to polyvanadates occur: at pH 4-6 $[H_2VO_4]^-$ is predominant at pV greater than ca. 4, while at higher concentrations trimers and tetramers are formed. Between pH 2-4 decavanadate predominates, its formation from orthovanadate is represented by this condensation reaction:

$$10\ [VO_4]^{3-} + 24\ H^+ \rightarrow [V_{10}O_{28}]^{6-} + 12\ H_2O$$

In decavanadate, each V (V) center is surrounded by six oxide ligands. Vanadic acid, H_3VO_4 exists only a very low concentrations because protonation of the tetrahedral species $[H_2VO_4]^-$ results in the preferential formation of the octahedral $[VO_2(H_2O)_4]^+$ species. In strongly acidic solutions, pH<2, $[VO_2(H_2O)_4]^+$ is the predominant species, while the oxide V_2O_5 precipitates from solution at high concentrations.

Coloured complexes of vanadium

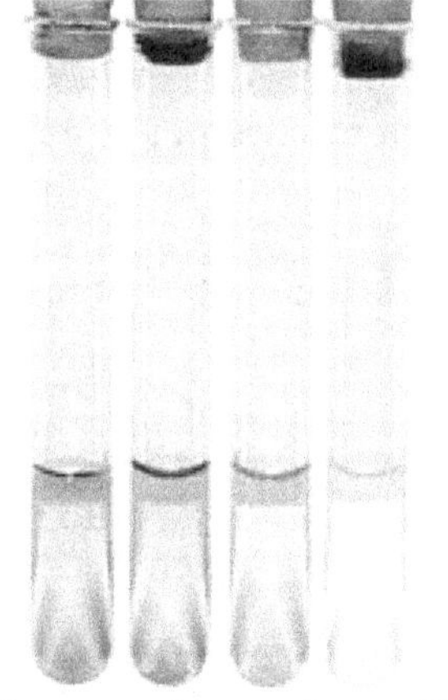

Fig. 35. Colours of vanadium metal complexes

The chemistry of vanadium is noteworthy for the accessibility of the four adjacent oxidation states 2-5. In aqueous solution, vanadium forms metal aquo complexes the colours (Fig. 35) are lilac $[V(H_2O)_6]^{2+}$, green $[V(H_2O)_6]^{3+}$, blue $[VO(H_2O)_5]^{2+}$, yellow VO_3^-. Vanadium (II) compounds are reducing agents, and vanadium (V) compounds are oxidizing agents. Vanadium (IV) compounds often exist as vanadyl derivatives which contain the VO^{2+} center.

Ammonium vanadate (V) (NH_4VO_3) can be successively reduced with elemental zinc to obtain the different colors of vanadium in these four oxidation states. Lower oxidation states occur in compounds such as

$V(CO)_6$, $[V(CO)_6]^-$ and substituted derivatives.

Halides of vanadium

Twelve binary halides, compounds with the formula VX_n, are known. VI_4, VCl_5, VBr_5, and VI_5 do not exist or are extremely unstable. In combination with other reagents, VCl_4 is used as a catalyst for polymerization of dienes. Like all binary halides, those of vanadium are Lewis acidic, especially those of V(IV) and V(V). Many of the halides form octahedral complexes with the formula VX_nL_{6-n} (X = halide; L = other ligand).Many vanadium oxyhalides (formula VO_mX_n) are known. The oxytrichloride and oxytrifluoride, (VOF_3) and $VOCl_3$) are the most widely studied. Akin to $POCl_3$, they are volatile, adopt tetrahedral structures in the gas phase, and are Lewis acidic.

4.3.4. Chromium:

Chromium is a chemical element which has the symbol **Cr** and atomic number 24. It is the first element in Group 6. The name of the element is derived from the Greek word "chrōma" meaning colour, because many of its compounds are intensely coloured.

Fig. 36. Chromium metal

- **Occurrence of chromium**

Fig. 37. Ruby containing chromium

Chromium is the 24th most abundant element in Earth's crust with an average concentration of 100 ppm. Chromium is mined as **chromite** **($FeCr_2O_4$)** ore. Chromium compounds are found in the environment, due to erosion of chromium-containing rocks and can be distributed by volcanic eruptions. The red colour (Fig.37) of the ruby shown is due to chromium.

- **Extraction of chromium**

Chromium is extracted from the chromite ore, $FeCr_2O_4$. An iron-chromium alloy, ferrochrome, is made by reducing chromite ore with carbon:

$$FeCr_2O_4 + 4C \longrightarrow [Fe + 2\ Cr] + 4CO\ (g)$$

↓ Ferrochrome

Ferrochrome and other alloying elements can be added directly to iron to produce steel. When needed, pure chromium can be prepared by reducing Cr_2O_3 in the thermite reaction.

$$Cr_2O_3(s) + 2\ Al(s) \longrightarrow Al_2O_3(s) + 2\ Cr.$$

- ## **Properties of chromium**

Chromium metal has proven of high value due to its high corrosion resistance and hardness. A major development was the discovery that steel could be made highly resistant to corrosion and discoloration by adding metallic chromium to form stainless steel. These applications, along with chrome plating (electroplating with chromium) currently comprise 85% of the commercial use for the element, with applications for chromium compounds forming the remainder.

Chromium is remarkable for its magnetic properties: it is the only elemental solid which shows antiferromagnetic ordering at room temperature. Above 38 °C, it transforms into a paramagnetic state. Chromium metal left standing in air is passivated by oxygen, forming a thin protective oxide surface layer. This passive oxide layer acts as protective layer to chromium. Passivated chromium is stable against acids.

Chromium, unlike metals such as iron and nickel, does not suffer from hydrogen embrittlement. However, it does suffer from nitrogen embrittlement, reacting with nitrogen from air and forming brittle nitrides at the high temperatures necessary to work the metal parts.

- ## **Applications of chromium**

- ➤ Chromium is an alloying element for steel. Chromium imparts high corrosion resistance and strength to steel. The high-speed tool steels contain between 3 and 5% chromium. Stainless steel, the main corrosion-proof metal alloy, is formed when chromium is added to iron in sufficient concentrations, usually above 11%.

- ➤ Chromium is also used in making nickel-based alloys. For example, Inconel 718 contains 18.6% chromium. Because of the excellent high-temperature properties of these nickel superalloys, they are used in jet engines and gas turbines in lieu of common structural materials.

- ➤ The relative high hardness and corrosion resistance of unalloyed chromium makes it a good surface coating. This is called chrome plating.

- ➤ In the chromate conversion coating process, the strong oxidative properties of chromates are used to deposit a protective oxide layer on metals like aluminium, zinc and cadmium.

- ➤ Anodizing of aluminium is another electrochemical process, which does not lead to the deposition of chromium, but uses chromic acid as electrolyte in the solution. During anodization, an oxide layer is formed on the aluminium.

- ➤ Chromium is an essential trace element in mammalian metabolism. In addition to insulin, it is responsible for reducing blood glucose levels, and is used to control certain cases of diabetes.

- ➤ It has also been found to reduce blood cholesterol levels by diminishing the concentration of (bad) low density lipoproteins "LDLs" in the blood. It is supplied in a variety of foods such as Brewer's yeast, liver, cheese, whole grain breads and cereals, and broccoli. It is claimed to aid in muscle development, and as such dietary supplements containing chromium picolinate (its most soluble form), is very popular with body builders.

- ### Compounds of chromium

Most compounds of chromium are coloured (why is $Cr(CO)_6$ white?); the most important are the chromates and dichromates of sodium and potassium and ammonium chrome alums. The dichromates are used as oxidizing agents in quantitative analysis, also in tanning leather. Other compounds are of industrial value; lead chromate is chrome yellow, a valued pigment. Chromium compounds are used in the textile industry as mordants, and by the aircraft and other industries for anodizing aluminium.

Oxides of chromium

The main oxides of chromium are Cr_2O_3 and CrO_3. Both Cr_2O_3 and the corresponding hydroxide, $Cr(OH)_3$ are amphoteric. $Cr(OH)_3$ reacts with nitric acid to produce violet coloured $[Cr(H_2O)_6]^{3+}$ and also reacts with base NaOH to give gray green coloured $[Cr(OH)_4]^-$.

The oxide Cr_2O_3, in contrast, has only acidic properties. It dissolves in water to produce a strongly acidic solution,

$$2\ Cr_2O_3(s) + 3\ H_2O\ (l) \rightarrow 2H_3O^+\ (aq) + Cr_2O_7^{2-}\ (aq)$$

CrO_3 has low melting point, since it has a high degree of covalency. Forms corner sharing tetrahedral.

Chromium (VI) species

Chromium (VI) compounds are powerful oxidants at low or neutral pH. Most important are chromate anion (CrO_4^{2-}) and dichromate ($Cr_2O_7^{2-}$) anions, which exist in equilibrium:

$$2\ [CrO_4]^{2-} + 2\ H^+ \rightleftharpoons [Cr_2O_7]^{2-} + H_2O$$

Chromium (VI) halides are known also and include the hexafluoride CrF_6 and chromyl chloride (CrO_2Cl_2). Sodium chromate is produced industrially by the oxidative roasting of chromite ore with calcium or sodium carbonate. The dominant species is therefore, by the law of mass action, determined by the pH of the solution. The change in equilibrium is visible by a change from yellow (chromate) to orange (dichromate), such as when an acid is added to a neutral solution of potassium chromate. At yet lower pH values, further condensation to more complex oxyanions of chromium is possible.

Both the chromate and dichromate anions are strong oxidizing reagents at low pH:

$$Cr_2O_7^{2-} + 14\ H_3O^+ + 6\ e^- \rightarrow 2\ Cr^{3+} + 21\ H_2O\ (\varepsilon_0 = 1.33\ V)$$

Dichromate anion is commonly used to oxidize alcohols to aldehydes, ketones and carboxylic acids. It is also extensively used in the analytical laboratory and its solutions are remarkably stable, remaining unchanged in concentration for months or even years.

They are, however, only moderately oxidizing at high pH:

$$CrO_4^{2-} + 4\,H_2O + 3\,e^- \rightarrow Cr(OH)_3 + 5\,OH^- (\varepsilon_0 = -0.13\ V)$$

Chromium (VI) compounds in solution can be detected by adding an acidic hydrogen peroxide solution. The unstable dark blue chromium (VI) peroxide (CrO_5) is formed, which can be stabilized as ether adduct $CrO_5\text{-}OR_2$.

Chromic acid has the hypothetical formula H_2CrO_4. It is a vaguely described chemical, despite many well-defined chromates and dichromates being known. The dark red chromium (VI) oxide CrO_3, the acid anhydride of chromic acid, is sold industrially as "chromic acid". It can be produced by mixing sulfuric acid with dichromate, and is a strong oxidizing agent.

Chromium (V) and chromium (IV)

The oxidation state +5 is only realized in few compounds but are intermediates in many reactions involving oxidations by chromate. The only binary compound is the volatile chromium (V) fluoride (CrF_5). This red solid has a melting point of 30 °C and a boiling point of 117 °C. It can be synthesized by treating chromium metal with fluorine at 400 °C and 200 bar pressure. The peroxochromate (V) is another example of the +5 oxidation state. Potassium peroxochromate ($K_3[Cr(O_2)_4]$) is made by reacting potassium chromate with hydrogen peroxide at low temperatures. This red brown compound is stable at room temperature but decomposes spontaneously at 150–170 °C.

Compounds of chromium (IV) (in the +4 oxidation state) are slightly more common than those of chromium(V). The tetrahalides, CrF_4, $CrCl_4$, and $CrBr_4$, can be produced by treating the trihalides ($CrX3$) with the corresponding halogen at elevated temperatures. Such compounds are susceptible to disproportionation reactions and are not stable in water.

<u>Chromium(III)</u>

CrX_3 (CrF_3 $CrCl_3$ $CrBr_3$ CrI_3) are prepared from Cr with X_2, dehydration of $CrCl_3.6H_2O$ requires $SOCl_2$ at 65^0C.

<u>Chromium(II)</u>

Many chromium (II) compounds are known, including the water-stable chromium(II) chloride, $CrCl_2$, which can be made by reduction of chromium(III) chloride with zinc. The resulting bright blue solution is only stable at neutral pH. Many chromous carboxylates are also known, most famously, the red chromous acetate ($Cr_2(O_2CCH_3)_4$), which features a quadruple bond.

Chromium(I)

Most Cr(I) compounds are obtained by oxidation of electron-rich, octahedral Cr(0) complexes. Other Cr(I) complexes contain cyclopentadienyl ligands. As verified by X-ray diffraction, a Cr-Cr quintuple bond (length 183.51(4) pm) has also been described. Extremely bulky monodentate ligands stabilize this compound by shielding the quintuple bond from further reactions.

Chromium compound determined experimentally to contain a Cr-Cr quintuple bond

Chromium(0)

Many chromium(0) compounds are known. Most are derivatives of chromium hexacarbonyl or bis(benzene)chromium.

4.3.5. Manganese:

Manganese is a chemical element, designated by the symbol **Mn** (Fig. 38). It has the atomic number 25. Manganese is a metal with important industrial metal alloy uses, particularly in stainless steels. In biology, manganese(II) ions function as cofactors for a large variety of enzymes with many functions. Manganese enzymes are particularly essential in detoxification of superoxide free radicals in organisms that must deal with elemental oxygen. Manganese also functions in the oxygen-evolving complex of photosynthetic plants.

Fig. 38. Manganese metal

- **Occurrence of manganese**

Fig. 39. Pyrolusite ore

Manganese is relatively abundant (0.1% of the earth crust-12th most abundant element). Manganese occurs principally as **pyrolusite (MnO_2-shown in Fig. 39)**, braunite, $(Mn^{2+}Mn^{3+}_6)(SiO_{12})$, psilomelane $(Ba,H_2O)_2Mn_5O_{10}$, and to a lesser extent as rhodochrosite ($MnCO_3$). The most important manganese ore is pyrolusite (MnO_2).

- **Extraction of manganese**

Manganese is obtained mainly from the mineral pyrolusite, MnO_2. Relatively pure manganese can be made by reducing MnO_2. However because most manganese is used in steel alloys, common practice is to reduce a mixture of MnO_2 and Fe_2O_3 to obtain an iron-manganese alloy called ferromanganese.

$$MnO_2 + Fe_2O_3 + 5C \longrightarrow [\underline{Mn + 2Fe}] + 5CO(g)$$

Ferromanganese

- ## <u>Properties of manganese</u>

Manganese is a silvery-gray metal that resembles iron. It is hard and very brittle, difficult to fuse, but easy to oxidize. Manganese metal and its common ions are paramagnetic. Manganese tarnishes slowly in air and "rusts" like iron, in water containing dissolved oxygen. Highly electropositive compared to its neighbours in Periodic Table. Mn is More reactive and It Will oxidise in air but not on a large scale. Burns in air if finely divided. Dissolves in acid to form Mn(II) salts. Not particularly reactive at room temperature, but will react with most things after heating. +2 state very stable (high spin d^5 – symmetrical). This breaks the trend – more resistant to oxidation than either Cr of Fe. Potassium permanganate, sodium permanganate and barium permanganate are all potent oxidizers. Potassium permanganate, also called Condy's crystals, is a commonly used laboratory reagent because of its oxidizing properties and finds use as a topical medicine (for example, in the treatment of fish diseases). Solutions of potassium permanganate were among the first stains and fixatives to be used in the preparation of biological cells and tissues for electron microscopy.

- ## <u>Applications of manganese</u>

- ➢ Manganese is essential to iron and steel production by virtue of its sulfur-fixing, deoxidizing, and alloying properties. Manganese is a key component of low-cost stainless steel formulations.
- ➢ Ferromanganese alloys are wear resistant and shock resistant and are used for rail road tracks, bulldozers and road scrapers.
- ➢ Steel with 12% manganese was used for British steel helmets. This steel composition was discovered in 1882 by Robert Hadfield and is still known as Hadfield steel.

- ➢ Manganese phosphating is used as a treatment for rust and corrosion prevention on steel.
- ➢ Manganese ions have various colors and are used industrially as pigments.
- ➢ Manganese dioxide is used as the cathode (electron acceptor) material in zinc-carbon and alkaline batteries.
- ## <u>Compounds of manganese</u>

<u>Oxides</u>

Mn_2O_7 requires prior oxidation to +7 state. MnO_2 is the most useful. Corner-sharing MnO_4 tetrahedra with Mn-O-Mn bridging. Not the most stable though, it will decompose to Mn_2O_3.

Mn_3O_4 has spinel structure with some Jahn-Teller distortion (Mn(III)). MnO can be made by reduction of any of the above with H_2. Basic oxide- Rock salt structure.

114

Halides

MnF4 is highest halide for Manganese (note that this shows Mn is less able to reach high oxidation states than Chromium). MnF3 forms octahedra which distort due to Jahn-Teller (d4). Oxohalides are explosively unstable.

Complexes

Many d-electrons available for back-donation from metal to ligand, stabilising low oxidation states. Oxidation State VII, VI & V – not really any. Oxidation State IV highest for stable complexes. Mostly dimeric or polymeric (as in chlorophyll, photosynthesis). Oxidation State III – high spin d^4. Small Jahn-Teller distortions also observed. Mn(III) is strongly oxidising and tends to disproportionate, but is stabilised by O-donor ligands. $[Mn(CN)_6]^{3-}$ is low spin however (good ligand – π bonding, covalency. Easily oxidised). Oxidation State II – $MnSO_4$ very important complex. Surprising thermal stability (compared to Fe, Co and Ni which all decompose quite easily under heat). In water this forms the aquated ion with high spin (t_2g^3 eg^2). d^5 configuration gives no CFSE so stability constants are low compared to other 1st row M(II) ions. Also indicates why a wide range of stereochemistry is observed for these complexes (no really advantage to any of them if no CFSE).

Potassium permanganate- KMnO$_4$

Potassium permanganate is an important oxidizing agent that is used in both analytical and organic chemistry laboratories. Its oxidizing power also makes it useful as a medical disinfectant and as a substitute for Cl_2 in water purification. In oxidation-reduction reactions in acidic solutions, MnO_4^- is usually reduced to Mn^{2+}. In basic solutions, it is reduced to MnO_2. Potassium permanganate is a self indicator.

4.3.6. Iron:

Iron is a chemical element with the symbol **Fe** (from Latin: *ferrum*) and atomic number 26. Fe is a reactive metal and is not found in free state in nature. It is the fourth most common element in the Earth's crust. Iron find very significant role as structural material after alloying to give Steel. Steels and low carbon iron alloys with other metals (alloy steels) are by far the most common metals in industrial use, due to their great range of desirable properties and the abundance of iron.

Fig. 40. Iron metal

- **Occurrence of Iron**

Iron is the sixth most abundant element in the Universe, and the most common refractory element. It is formed as the final exothermic stage of stellar nucleosynthesis, by silicon fusion in massive stars. Fe ores are in the form of compounds of oxides, sulphides and carbonates. The ores found are:

- **Hematite: Fe_2O_3**   • Magnetite: Fe_3O_4 • Limonite: $Fe_2O_3. H_2O$

- Siderite: $FeCO_3$ • Pyrite: FeS_2

Metallic or native iron is rarely found on the surface of the Earth because it tends to oxidize, but its oxides are pervasive and represent the primary ores. While it makes up about 5% of the Earth's crust, both the Earth's inner and outer core are believed to consist largely of an iron-nickel alloy constituting 35% of the mass of the Earth as a whole. Iron is consequently the most abundant element on Earth, but only the fourth most abundant element in the Earth's crust. Most of the iron in the crust is found combined with oxygen as iron oxide minerals such as hematite and magnetite. Large deposits of iron are found in banded iron formations. These geological formations are a type of rock consisting of repeated thin layers of iron oxides, either magnetite (Fe_3O_4) or hematite (Fe_2O_3), alternating with bands of iron-poor shale and chert.

- ## Extraction of Iron

Iron is extracted as cast iron mostly from hematite ore. The extraction is carried out in the following steps:

1. **Concentration**: The ore is crushed into small pieces of about 2.5-5.0 cm size. In most of the cases, the ore is rich enough and does not require any further concentration. However in few cases, the ore is washed with a stream of water and if necessary, concentrated by magnetic separation technique.

2. **Calcination**: The concentrated ore is heated strongly in the presence of air in large open heaps or in special kilns. During calcinations, following changes take place:

i) Moisture is removed and the ore gets dried.

ii) The impurities of sulphur, arsenic, phosphorus etc. get oxidized and are removed in the form of their volatile oxides.

$$S + O_2 \longrightarrow SO_2 \uparrow$$

$$4As + 3O_2 \longrightarrow 2As_2O_3 \uparrow$$

$$4P + 5O_2 \longrightarrow 2P_2O_5 \uparrow$$

iii) Hydrated ferric oxide if present gets dehydrated while $FeCO_3$ decomposes into FeO.

$$Fe_2O_3.3H_2O \longrightarrow Fe_2O_3 + 3H_2O$$

$$FeCO_3 \longrightarrow FeO + CO_2$$

iv) The ore becomes porous. This helps in its reduction at a later stage.

3. **Smelting**: The calcined ore is mixed with lime stone and coke and is fed to a blast furnace. In this furnace the ore is subjected to smelting.

Blast Furnace

Blast furnace consists of chimney like structure [24-30 M high] made of iron or steel and lined with firebricks. It is narrow at the top, increasing the diameter downward to form the crucible. Blast furnace has,

a) *Cup and cone arrangement* at the top for the introduction of charge. It has an outlet for the exit of waste gases. It prevents the exit of hot gases during the charging process.

b) *Tuyers* are water jacketed pipes through which hot air is blown in to the furnace where combustion takes place.

c) *Hearth or Crucible*: This is lowest part of the furnace. The hearth serves as a large crucible which has tapping hole for removing molten iron and an outlet for the slag.

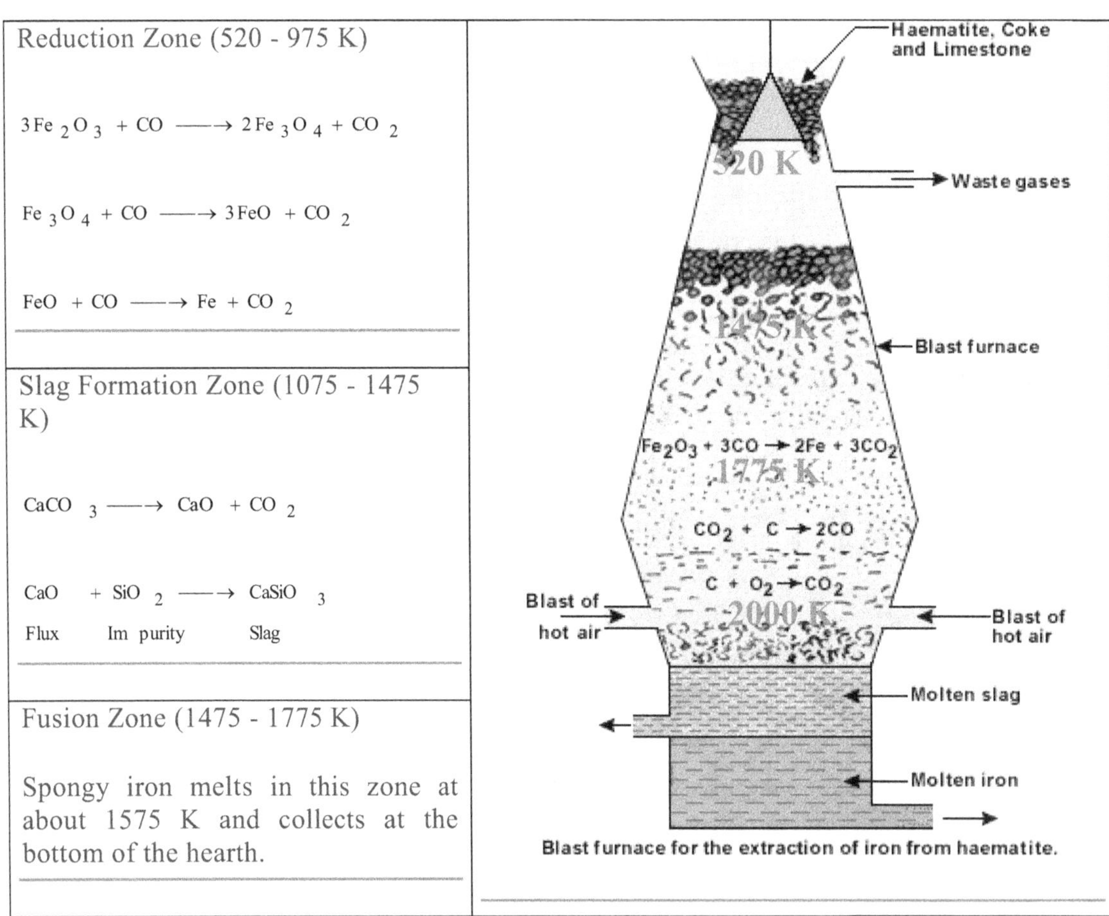

Reduction Zone (520 - 975 K)

$$3Fe_2O_3 + CO \longrightarrow 2Fe_3O_4 + CO_2$$

$$Fe_3O_4 + CO \longrightarrow 3FeO + CO_2$$

$$FeO + CO \longrightarrow Fe + CO_2$$

Slag Formation Zone (1075 - 1475 K)

$$CaCO_3 \longrightarrow CaO + CO_2$$

$$CaO + SiO_2 \longrightarrow CaSiO_3$$

Flux   Im purity   Slag

Fusion Zone (1475 - 1775 K)

Spongy iron melts in this zone at about 1575 K and collects at the bottom of the hearth.

Blast furnace for the extraction of iron from haematite.

Fig. 40. The various zones of Blast furnace

The charge introduced from the top of the furnace consists of calcinated ore. Coke and lime [8:4:1]

- Coke serves the purpose of fuel and reducing agent.
- Lime acts as flux

Coke undergo combustion to heat the furnace to different temperatures in different zones of it. Iron thus obtained from the blast furnace is called Pig iron. It contains about 3-5% carbon and varying amounts of S, Si, P and Mn. Pig iron is remelted in a vertical furnace heated by coke and the molten metal is poured into moulds. Iron thus obtained is called Cast iron.

- ***Pig iron*** *has 3.5–4.5% carbon and contains varying amounts of contaminants such as sulfur, silicon and phosphorus. Pig iron is not a saleable product, but rather an intermediate step in the production of cast iron and steel.*

- ***Wrought*** *iron contains less than 0.25% carbon but large amounts of slag that give it a fibrous characteristic. It is a tough, malleable product, but not as fusible as pig iron. If honed to an edge, it loses it quickly. Wrought iron is characterized by the presence of fine fibers of slag entrapped within the metal. Wrought iron is more corrosion resistant than steel. It has been almost completely replaced by mild steel for traditional "wrought iron" products and blacksmithing.*

- *In **gray iron** the carbon exists as separate, fine flakes of graphite, and also renders the material brittle due to the sharp edged flakes of graphite that produce stress concentration sites within the material.*

- ## Properties of Iron

Elemental iron occurs in meteoroids and other low oxygen environments, but is reactive to oxygen and water. Fresh iron surfaces appear lustrous silvery-gray, but oxidize in normal air to give hydrated iron oxides, commonly known as rust. Unlike many other metals which form passivating oxide layers, iron oxides occupy more volume than iron metal, and thus iron oxides flake off and expose fresh surfaces for corrosion.

Iron metal has been used since ancient times, though copper alloys, which have lower melting temperatures, were used first in history. Pure iron is soft (softer than aluminium), but is unobtainable by smelting. The material is significantly hardened and strengthened by impurities from the smelting process, such as carbon. A certain proportion of carbon (between 0.002% and 2.1%) produces steel, which may be up to 1000 times harder than pure iron. Iron chemical compounds, which include ferrous and ferric compounds, have many uses.

Iron oxide mixed with aluminium powder can be ignited to create a thermite reaction, used in welding and purifying ores. It forms binary compounds with the halogens and the chalcogens. Among its organometallic compounds is ferrocene, the first sandwich compound discovered.

Iron plays an important role in biology, forming complexes with molecular oxygen in hemoglobin and myoglobin; these two compounds are common oxygen transport proteins in vertebrates. Iron is also the metal used at the active site of many important redox enzymes dealing with cellular respiration and oxidation and reduction in plants and animals.

- **<u>Applications of Iron</u>**

Iron is the most widely used of all the metals, accounting for 95% of worldwide metal production.

- ➢ Its low cost and high strength make it indispensable in engineering applications such as the construction of machinery and machine tools, automobiles, the hulls of large ships, and structural components for buildings. Since pure iron is quite soft, it is most commonly combined with alloying elements to make steel.
- ➢ Iron is also used for protection from ionizing radiation. Although it is lighter than another traditional protection material, lead, it is much stronger mechanically.
- ➢ Iron catalysts are traditionally used in the Haber-Bosch Process for the production of ammonia and the Fischer-Tropsch process for conversion of carbon monoxide to hydrocarbons for fuels and lubricants.
- ➢ Powdered iron in an acidic solvent was used in the Bechamp reduction the reduction of nitrobenzene to aniline.
- ➢ Iron(III) chloride finds use in water purification and sewage treatment, in the dyeing of cloth, as a coloring agent in paints, as an additive in animal feed, and as an etchant for copper in the manufacture of printed circuit boards.
- ➢ It can also be dissolved in alcohol to form tincture of iron.
- ➢ Iron (II) sulfate is used as a precursor to other iron compounds. It is also used to reduce chromate in cement. It is used to fortify foods and treat iron deficiency anemia. These are its main uses.

- ➢ Iron (II) chloride is used as a reducing flocculating agent, in the formation of iron complexes and magnetic iron oxides, and as a reducing agent in organic synthesis.

- ➢ The most commonly known and studied "bioinorganic" compounds of iron (i.e., iron compounds used in biology) are the heme proteins: examples are hemoglobin, myoglobin, and cytochrome. These compounds can transport gases, build enzymes, and be used in transferring electrons. Metalloproteins are a group of proteins with metal ion cofactors. Some examples of iron metalloproteins are ferritin and rubredoxin. Many enzymes vital to life contain iron, such as catalase, lipoxygenases, and IRE-BP

- ## **Compounds of iron**

Ferrous ammonium sulphate

Ammonium iron (II) sulfate, or **Mohr's Salt**, is the inorganic compound with the formula $(NH_4)_2Fe(SO_4)_2 \cdot 6H_2O$.. Containing two different cations, Fe^{2+} and NH_4^+, it is classified as a double salt of ferrous sulfate and ammonium sulfate. It is a common laboratory reagent. Like the other ferrous sulfate salts, ferrous ammonium sulfate dissolves in water to give the aquo complex $[Fe(H_2O)_6]^{2+}$, which has octahedral molecular geometry. It acts as very good reducing agents in acidic medium. The reaction is as shown below.

$$K_2Cr_2O_7 + 6\,FeSO_4 + 7\,H_2SO_4 \longrightarrow K_2SO_4 + Cr_2(SO_4)_3 + 3\,Fe_2(SO_4)_3 + 7\,H_2O$$

Orange   *Green*

In analytical chemistry, this salt is preferred over other salts of ferrous sulfate for titration purposes as it is much less prone to oxidation by air to iron (III). The oxidation of solutions of iron (II) is very pH dependent, occurring much more readily at high pH. The ammonium ions make solutions of Mohr's salt slightly acidic, which slows this oxidation process. Sulfuric acid is commonly added to solutions to reduce oxidation to ferric iron.

Mohr's salt is named after the German chemist Karl Friedrich Mohr, who made many important advances in the methodology of titration in the 19th century.

Iron oxides

Iron oxides are chemical compounds composed of iron and oxygen. All together, there are sixteen known iron oxides and oxyhydroxides. Iron oxides and oxide-hydroxides are widespread in nature, play an important role in many geological and biological processes. Common rust is a form of iron(III) oxide. Iron oxides are widely used as inexpensive, durable pigments in paints, coatings and colored concretes. Colors commonly available are in the "earthy" end of the yellow/orange/red/brown/black range.

- iron(II) oxide, wüstite (FeO)
- iron(II,III) oxide, magnetite (Fe_3O_4)
- iron(III) oxide (Fe_2O_3)
    - alpha phase, hematite (α-Fe_2O_3)
    - beta phase, (β-Fe_2O_3)
    - gamma phase, maghemite (γ-Fe_2O_3)
    - epsilon phase, (ε-Fe_2O_3)

Iron(III) oxide-hydroxide

A number of chemicals are dubbed iron(III) oxide-hydroxide. These chemicals are oxide-hydroxides of iron, and may occur in anhydrous (FeO(OH)) or hydrated (FeO(OH)·nH_2O) forms. The monohydrate (FeO(OH)·H_2O) might otherwise be described as iron(III) hydroxide ($Fe(OH)_3$), and is also known as hydrated iron oxide or yellow iron oxide. Yellow iron oxide (CAS [51274-00-1]) is used as a pigment, e.g. Pigment Yellow 42. Pigment Yellow 42 is Food and Drug Administration (FDA) approved for use in cosmetics and is used in some tattoo inks.

Prussian blue and Turnbull's blue

Prussian blue, $\{[Fe_4[Fe(CN)_6]_3\}$ is a dark blue pigment with the idealized formula $Fe_7(CN)_{18}$. Another name for the color *Prussian blue* is *Berlin blue* or, in painting, *Parisian blue*. **Turnbull's blue** is the same substance but is made from different reagents, and its slightly different color stems from different impurities.

Prussian blue was one of the first synthetic pigments. It is employed as a very fine colloidal dispersion, as the compound itself is not soluble in water.A "soluble" form of PB, $K[Fe^{III}Fe^{II}(CN)_6]$, which is really colloidal, can be made from potassium ferrocyanide and iron(III): $\quad K^+ + Fe^{3+} + [Fe^{II}(CN)_6]^{4-} \rightarrow KFe^{III}[Fe^{II}(CN)_6]$

The similar reaction of potassium ferricyanide and iron(II) results in the same colloidal solution, because $[Fe^{III}(CN)_6]^{3-}$ is converted into ferrocyanide. Prussian blue is produced if in the reactions above an excess of Fe^{3+} or Fe^{2+}, respectively, is added. In the first case:

$$4Fe^{3+} + 3[Fe^{II}(CN)_6]^{4-} \rightarrow \mathbf{Fe^{III}[Fe^{III}Fe^{II}(CN)_6]_3}$$

Prussian blue, the pigment, is used in paints, and it is the traditional "blue" in blueprints. In medicine, Prussian blue is used as an antidote for certain kinds of heavy metal poisoning, e.g., by caesium and thallium. In particular it was used to absorb $^{137}Cs^+$ from those poisoned in the Goiânia accident. Prussian blue is orally administered. The therapy exploits Prussian Blue's ion exchange properties and high affinity for certain "soft" metal cations.

4.3.7. Cobalt:

Cobalt is a chemical element with symbol **Co (Fig. 41)** and atomic number 27. Like nickel, cobalt in the Earth's crust is found only in chemically combined form, save for small deposits found in alloys of natural meteoric iron. The free element, produced by reductive smelting, is a hard, lustrous, silver-gray metal. Miners had long used the name *kobold ore* (German for *goblin ore*) for some of the blue-pigment producing minerals; they were named because they were poor in known metals, and

Fig. 41. Cobalt metal

gave poisonous arsenic-containing fumes upon smelting. In 1735, such ores were found to be reducible to a new metal (the first discovered since ancient times), and this was ultimately named for the *kobold*.

- **Occurrence of cobalt**

Free cobalt (the native metal) is not found in on Earth. Though the element is of medium abundance, natural compounds of cobalt are numerous. Small amounts of cobalt compounds are found in most rocks, soil, plants, and animals. It is the major metallic component in combination with sulfur and arsenic in the sulfidic **cobaltite (CoAsS)**, safflorite ($CoAs_2$), glaucodot ($(Co,Fe)AsS$), and skutterudite ($CoAs_3$) minerals(Cobalt ore shown in Fig. 42).

Fig. 42. Cobalt ore

- **Extraction of cobalt**

Cobalt is not obtained by active mining of cobalt ores, but rather by reducing cobalt compounds that occur as by-products of nickel and copper mining activities. Several methods exist for the separation of cobalt from copper and nickel. They depend on the concentration of cobalt and the exact composition of the used ore. One separation step involves froth flotation, in which surfactants bind to different ore components, leading to an enrichment of cobalt ores. Subsequent roasting converts the ores to the cobalt sulfate, whereas the copper and the iron are oxidized to the oxide. The leaching with water extracts the sulfate together with the arsenates. The residues are further leached with sulfuric acid yielding a solution of copper sulfate. Cobalt can also be leached from the slag of the copper smelter.

The products of the above-mentioned processes are transformed into the cobalt oxide (Co_3O_4). This oxide is reduced to the metal by the aluminothermic reaction or reduction with carbon in a blast furnace.

- **Properties of cobalt**

Cobalt is a ferromagnetic metal with a specific gravity of 8.9. The Curie temperature is 1115 °C and the magnetic moment is 1.6–1.7 Bohr magnetons per atom. Cobalt has a relative permeability two thirds that of iron. Metallic cobalt occurs as two crystallographic structures: hcp and fcc. The ideal transition temperature between the hcp and fcc structures is 450 °C, but in practice, the energy difference is so small that random intergrowth of the two is common.

Cobalt is a weakly reducing metal that is protected from oxidation by a passivating oxide film. It is attacked by halogens and sulfur. Heating in oxygen produces Co_3O_4 which loses oxygen at 900 °C to give the monoxide CoO. The metal reacts with Fluorine gas (F_2) at 520 K to give CoF_3; with chlorine (Cl_2), bromine (Br_2) and iodine (I_2), the corresponding binary halides are formed. It does not react with hydrogen gas (H_2) or nitrogen gas (N_2) even when heated, but it does react with boron, carbon, phosphorus, arsenic and sulphur. At ordinary temperatures, it reacts slowly with mineral acids, and very slowly with moist, but not with dry, air.

- **Applications of cobalt**

  ➢ The main application of cobalt is as the free metal, in production of certain high performance super alloys suitable for use in turbine blades for gas turbines and jet aircraft engines.

- Cobalt-based alloys are also corrosion and wear-resistant. This makes them useful in the medical field, where cobalt is often used (along with titanium) for orthopedic implants that do not wear down over time.
- Special cobalt-chromium-molybdenum alloys like Vitallium are used for prosthetic parts such as hip and knee replacements.
- Cobalt alloys are also used for dental prosthetics, where they are useful to avoid allergies to nickel
- The special alloys of aluminium, nickel, cobalt and iron, known as Alnico, and of samarium and cobalt (samarium-cobalt magnet) are used in permanent magnets.
- It is also alloyed with 95% platinum for jewelry purposes, yielding an alloy that is suitable for fine detailed casting and is also slightly magnetic.
- Lithium cobalt oxide ($LiCoO_2$) is widely used in lithium ion battery cathodes. The material is composed of cobalt oxide layers in which the lithium is intercalated.
- Several cobalt compounds are used in chemical reactions as oxidation catalysts. Cobalt acetate is used for the conversion of xylene to terephthalic acid, the precursor to the bulk polymer polyethylene terephthalate. Typical catalysts are the cobalt carboxylates (known as cobalt soaps). They are also used in paints, varnishes, and inks as "drying agents" through the oxidation of drying oils.
- Cobalt-57 (Co-57 or ^{57}Co) is a cobalt radioisotope most often used in medical tests, as a radiolabel for vitamin B_{12} uptake.
- The two varieties of cobalt blue pigment, cobalt blue (cobalt aluminate) and cobalt green (a mixture of cobalt(II) oxide and zinc oxide), were used as pigments for paintings because of their superior stability.
- Cobalt is essential to all animals. It is a key constituent of cobalamin, also known as vitamin B_{12}, which is the primary biological reservoir of cobalt as an "ultratrace" element.

- **<u>Compounds of cobalt</u>**

<u>**Oxygen and chalcogen compounds**</u>

Several oxides of cobalt are known. Green cobalt(II) oxide (CoO) has rocksalt structure. It is readily oxidized with water and oxygen to brown cobalt(III) hydroxide ($Co(OH)_3$). At temperatures of 600–700 °C, CoO oxidizes to the blue cobalt(II,III) oxide (Co_3O_4), which has a spinel structure. Black cobalt(III) oxide (Co_2O_3) is also known. Cobalt oxides are antiferromagnetic at low temperature: CoO and Co_3O_4, which is analogous to magnetite (Fe_3O_4), with a mixture of +2 and +3 oxidation states.

The principal chalcogenides of cobalt include the black cobalt(II) sulfides, CoS_2, which adopts a pyrite-like structure, and cobalt(III) sulfide (Co_2S_3). Pentlandite (Co_9S_8) is metal-rich.

<u>**Halides**</u>

Four dihalides of cobalt(II) are known: cobalt(II) fluoride (CoF_2, pink), cobalt(II) chloride ($CoCl_2$, blue), cobalt(II) bromide ($CoBr_2$, green), cobalt(II) iodide (CoI_2, blue-black). These halides exist in anhydrous and hydrated forms. Whereas the anhydrous dichloride is blue, the hydrate is red. The reduction potential for the reaction $Co^{3+} + e^- \rightarrow Co^{2+}$

is +1.92 V, beyond that for chlorine to chloride, +1.36 V. As a consequence cobalt(III) and chloride would result in the cobalt(III) being reduced to cobalt(II). Because the reduction potential for fluorine to fluoride is so high, +2.87 V, cobalt(III) fluoride is one of the few simple stable cobalt(III) compounds. Cobalt(III) fluoride, which is used in some fluorination reactions, reacts vigorously with water.

Coordination compounds

Alfred Werner, a Nobel-prize winning pioneer in coordination chemistry, worked with compounds of empirical formula $CoCl_3(NH_3)_6$. One of the isomers determined was cobalt(III) hexammine chloride. This coordination complex, a "typical" Werner-type complex, consists of a central cobalt atom coordinated by six ammine ligands orthogonal to each other and three chloride counteranions. Using chelating ethylenediamine ligands in place of ammonia gives tris(ethylenediamine)cobalt(III) chloride ($[Co(en)_3]Cl_3$), which was one of the first coordination complexes that was resolved into optical isomers. The complex exists as both either right- or left-handed forms of a "three-bladed propeller". This complex was first isolated by Werner as yellow-gold needle-like crystals.

Organometallic compounds

Cobaltocene is a structural analog to ferrocene, where cobalt substitutes for iron. Cobaltocene is sensitive to oxidation, much more than ferrocene. Cobalt carbonyl ($Co_2(CO)_8$) is a catalyst in carbonylation reactions. Vitamin B_{12} is an organometallic compound found in nature and is the only vitamin to contain a metal atom.

4.3.8. Nickel:

Nickel is a chemical element with the chemical symbol **Ni** and atomic number 28. It is a silvery-white lustrous metal with a slight golden tinge. Nickel belongs to the transition metals and is hard and ductile. Pure nickel shows a significant chemical activity that can be observed when nickel is powdered to maximize the exposed surface area on which reactions can occur, but larger pieces of the metal are slow to react with air at ambient conditions due to the formation of a protective oxide surface. An iron–nickel mixture is thought to compose Earth's inner core.

Fig. 43. NiCkel metal

Occurrence of nickel

Nickel is the earth's 22nd most abundant element and the 7th most abundant transition metal. It is a silver white crystalline metal that occurs in meteors or combined with other elements in ores. Two important groups of ores are:

1. Laterites: oxide or silicate ores such as garnierite, $(Ni,Mg)_6 Si_4O_{10}(OH)_8$ which are predominantly found in tropical areas such as New Caledonia, Cuba and Queensland.

2. Sulphides: these are ores such as pentlandite, $(Ni,Fe)_9S_8$ which contain about 1.5%, nickel associated with copper, cobalt and other metals. They are predominant in more temperate regions such as Canada, Russia and South Africa.

Canada is the world's leading nickel producer and the Sudbury Basin of Ontario contains one of the largest nickel deposits in the world.

- ## Extraction of nickel

In 1899 Ludwig Mond developed a process for extracting and purifying nickel. The so-called "Mond Process" involves the conversion of nickel oxides to pure nickel metal. The oxide is obtained from nickel ores by a series of treatments including concentration, roasting and smelting of the minerals.
In the first step of the process, nickel oxide is reacted with water gas, a mixture of H_2 and CO, at atmospheric pressure and a temperature of 50 °C. The oxide is thus reduced to impure nickel. Reaction of this impure material with residual carbon monoxide gives the toxic and volatile compound, nickel tetracarbonyl, $Ni(CO)_4$. This compound decomposes on heating to about 230 °C to give pure nickel metal and CO, which can then be recycled.

The actual temperatures and pressures used in this process may vary slightly from one processing plant to the next. However the basic process as outlined is common to all.

The process can be summarised as follows:

$$\overset{50°C}{Ni + 4CO} \; \longrightarrow \; \underset{\text{(impure)}}{Ni(CO)_4} \; \overset{230°C}{\longrightarrow} \; \underset{\text{(pure)}}{Ni + 4CO}.$$

Properties of nickel

Nickel is a hard silver white metal, which occurs as cubic crystals. It is malleable, ductile and has superior strength and corrosion resistance. The metal is a fair conductor of heat and electricity and exhibits magnetic properties below 345°C. Five isotopes of nickel are known.

In its metallic form nickel is chemically unreactive. It is insoluble in cold and hot water and ammonia and is unaffected by concentrated nitric acid and alkalis. It is however soluble in dilute nitric acid and sparingly soluble in dilute hydrochloric and sulphuric acids.

- **Applications of nickel**
  - The primary use of nickel is in the preparation of alloys such as stainless steel, which accounts for approximately 67% of all nickel used in manufacture. The greatest application of stainless steel is in the manufacturing of kitchen sinks but it has numerous other uses as well.

  - An alloy of nickel and copper for example is a component of the tubing used in the desalination of sea water.
  - Nickel steel is used in the manufacture of armour plates and burglar proof vaults. Nickel alloys are especially valued for their strength, resistance to corrosion and in the case of stainless steel for example, aesthetic value.
  - Electroplating is another major use of the metal. Nickel plating is used in protective coating of other metals. In wire form, nickel is used in pins, staples, jewellry and surgical wire.
  - Finely divided nickel catalyses the hydrogenation of vegetable oils. Nickel is also used in the colouring of glass to which it gives a green hue.
  - Other applications of nickel include:
    -Coinage
    -Transportation and construction
    -Petroleum industry
    -Machinery and household appliances
    -Chemical industry.
  - Nickel compounds also have useful applications. Ceramics, paints and dyes, electroplating and preparation of other nickel compounds are all applications of these compounds. Nickel oxide for example is used in porcelain painting and in electrodes for fuel cells and batteries. Nickel acetate is used as a mordant in the textiles industry. Nickel carbonate finds use in ceramic colours and glazes.

- **Compounds of nickel**

Oxides/hydroxides of nickel

Nickel is known primarily for its divalent compounds since the most important oxidation state of the element is +2. There do exist however certain compounds in which the oxidation state of the metal is between -1 to +4. Nickel oxide is a powdery green solid that becomes yellow on heating. It is difficult to prepare this compound by simply heating nickel in oxygen and it is more conveniently obtained by heating nickel hydroxide, carbonate or nitrate. Nickel oxide is readily soluble in acids but insoluble in hot and cold swater. melting point, since it has a high degree of covalency. Forms corner sharing tetrahedral.

Thermal decomposition of $Ni(OH)_2$, $NiCO_3$, or $NiNO_3$ gives NiO.

Blue and green are the characteristic colours of nickel compounds and they are often hydrated. Nickel hydroxide usually occurs as green crystals that can be precipitated when aqueous alkali is added to a solution of a nickel (II) salt. It is insoluble in water but dissolves readily in acids and ammonium hydroxide.

Sulphides/Chlorides of nickel

Nickel sulfides consist of NiS_2, which has a pyrite structure, and Ni_3S_4, which has a spinel structure. All the nickel dihalides are known to exist. These compounds are usually yellow to dark brown in colour. Preparation directly from the elements is possible for all except NiF_2, which is best prepared from reaction of F_2 on $NiCl_2$ at 350°C. Most are soluble in water and crystallisation of the hexahydrate containing the $[Ni(H_2O)_6]^{2+}$ ion can be achieved. NiF_2 however is only slightly soluble in water from which the trihydrate crystallizes. The only nickel trihalide known to exist is an impure specimen of NiF_3.

Nickel(II) halides

Formula	Colour	MP	μ (BM)	Structure
NiF_2	yellow	1450	2.85	tetragonal rutile
$NiCl_2$	yellow	1001	3.32	$CdCl_2$
$NiBr_2$	yellow	965	3.0	$CdCl_2$
NiI_2	Black	780	3.25	$CdCl_2$

Identification of nickel compounds can be achieved by employing the use of an organic reagent dimethylglyoxine. This compound forms a red flocculent precipitate on addition to a solution of a nickel compound.

Nickel (0)

Tetracarbonylnickel ($Ni(CO)_4$), discovered by Ludwig Mond, is a volatile, highly toxic liquid at room temperature. On heating, the complex decomposes back to nickel and carbon monoxide:

$$Ni(CO)_4 \rightleftharpoons Ni + 4\ CO$$

This behavior is exploited in the Mond process for purifying nickel, as described above. The related nickel(0) complex bis(cyclooctadiene)nickel(0) is a useful catalyst in organonickel chemistry due to the easily displaced cod ligands.

Nickel (I)

Nickel (I) complexes are uncommon, one example being the tetrahedral complex $NiBr(PPh_3)_3$. Many feature Ni-Ni bonding, such as the dark red diamagnetic $K_4[Ni_2(CN)_6]$ prepared by reduction of $K_2[Ni_2(CN)_6]$ with sodium amalgam. This compound is oxidised in water, liberating H_2.

Nickel (II)

Nickel (II) forms compounds with all common anions, i.e. the sulfide, sulfate, carbonate, hydroxide, carboxylates, and halides. Nickel (II) sulfate is produced in large quantities by dissolving nickel metal or oxides in sulfuric acid. It exists as both a hexa- and heptahydrates. This compound is useful for electroplating nickel. Common salts of nickel, such as the chloride, nitrate, and sulfate, dissolve in water to give green solutions containing the metal aquo complex $[Ni(H_2O)_6]^{2+}$.

Some tetracoordinate nickel (II) complexes form both tetrahedral and square planar geometries. The tetrahedral complexes are paramagnetic whereas the square planar complexes are diamagnetic. This equilibrium as well as the formation of octahedral complexes contrasts with the behavior of the divalent complexes of the heavier group 10 metals, palladium (II) and platinum (II), which tend to adopt only square-planar geometry.

Nickel (III) and (IV)

For simple compounds, nickel (III) and nickel (IV) only occurs with fluoride and oxides, with the exception of $KNiIO_6$, which can be considered as a formal salt of the $[IO_6]^{5-}$ ion. Ni (IV) is present in the mixed oxide $BaNiO_3$, while Ni (III) is present in nickel (III) oxide, which is used as the cathode in many rechargeable batteries, including nickel-cadmium, nickel-iron, nickel hydrogen, and nickel-metal hydride, and used by certain manufacturers in Li-ion batteries Nickel (III) can be stabilized by σ-donor ligands such as thiols and phosphines.

4.3.9. Copper:

Copper is a chemical element with the symbol **Cu** (from Latin: *cuprum*) and atomic number 29. It is a ductile metal with very high thermal and electrical conductivity. Pure copper is soft and malleable; a freshly exposed surface has a reddish-orange color. It is used as a conductor of heat and electricity, a building material, and a constituent of various metal alloys. The metal and its alloys have been used for thousands of years. In the Roman era, copper was principally mined on Cyprus, hence the origin of the name of the metal as *cyprium* (metal of Cyprus), later shortened to *cuprum*.

Fig. 44. Copper metal

Copper is essential to all living organisms as a trace dietary mineral because it is a key constituent of the respiratory enzyme complex cytochrome c oxidase.

In molluscs and crustacea copper is a constituent of the blood pigment hemocyanin, which is replaced by the iron-complexed hemoglobin in fish and other vertebrates. The main areas where copper is found in humans are liver, muscle and bone. Copper compounds are used as bacteriostatic substances, fungicides, and wood preservatives.

Fig. 45. Copper Pendant

- ## Occurrence of copper

Chromium Copper is the earth's 25th most abundant element, but one of the less common first row transition metals. It occurs as a soft reddish metal that can be found native as large boulders weighing several hundred tons or as sulphide ores. The latter are complex copper, iron and sulphur mixtures in combination with other metals such as arsenic, zinc and silver. The copper concentration in such ores is typically between 0.5-2%.

The commonest ore is **chalcopyrite, $CuFeS_2$**, a brass yellow ore that accounts for approximately 50% of the world's copper deposits. Numerous other copper ores of varying colours and compositions exist. Examples are **malachite, $Cu_2CO_3(OH)_2$**, a bright green ore, and the **red ore cuprite, Cu_2O**. Humans first used copper about 10,000 years ago. A copper pendant (Shown in Fig. 45) discovered in Northern Iraq is thought to date back to around 8700 BC. Prehistoric man probably used copper for weapon making. Ancient Egyptians too seemed to have appreciated the corrosion resistance of the metal. They used copper bands and nails in ship building and copper pipes were used to convey water. Some of these artifacts survive today in good condition. An estimate of the total Egyptian copper output over 1500 years is 10,000 tons.

- ## Extraction of copper

The concentration of copper in ores averages only 0.6%, and most commercial ores are sulfides, especially chalcopyrite ($CuFeS_2$) and to a lesser extent chalcocite (Cu_2S). These minerals are concentrated from crushed ores to the level of 10–15% copper by froth flotation or bioleaching. Heating this material with silica in flash smelting removes much of the iron as slag. The process exploits the greater ease of converting iron sulfides into its oxides, which in turn react with the silica to form the silicate slag, which floats on top of the heated mass. The resulting *copper matte* consisting of Cu_2S is then roasted to convert all sulfides into oxides.

$$2\ Cu_2S + 3\ O_2 \rightarrow 2\ Cu_2O + 2\ SO_2$$

The cuprous oxide is converted to *blister* copper upon heating:

$$2\ Cu_2O \rightarrow 4\ Cu + O_2$$

The Sudbury matte process converted only half the sulfide to oxide and then used this oxide to remove the rest of the sulfur as oxide. It was then electrolytically refined and the anode mud exploited for the platinum and gold it contained. This step exploits the relatively easy reduction of copper oxides to copper metal. Natural gas is blown across the blister to remove most of the remaining oxygen and electrorefining is performed on the resulting material to produce pure copper:

$$Cu^{2+} + 2\ e^- \rightarrow Cu$$

Properties of copper

Copper, silver and gold are in group 11 of the periodic table, and they share certain attributes: they have one s-orbital electron on top of a filled d-electron shell and are characterized by high ductility and electrical conductivity. Pure copper is orange-red and acquires a reddish tarnish when exposed to air. Together with caesium and gold (both yellow), and osmium (bluish), copper is one of only four elemental metals with a natural color other than gray or silver. The characteristic color of copper results from the electronic transitions between the filled 3d and half-empty 4s atomic shells – the energy difference between these shells is such that it corresponds to orange light. The same mechanism accounts for the yellow color of gold and caesium.

Copper does not react with water, but it slowly reacts with atmospheric oxygen forming a layer of brown-black copper oxide. In contrast to the oxidation of iron by wet air, this oxide layer stops the further, bulk corrosion. A green layer of verdigris (copper carbonate) can often be seen on old copper constructions, such as the Statue of Liberty.

- ## Applications of copper

- ➢ The major applications of copper are in electrical wires (60%), roofing and plumbing (20%) and industrial machinery (15%).

- ➢ Electrical wiring is the most important market for the copper industry. This includes building wire, communications cable, power distribution cable, appliance wire, automotive wire and cable, and magnet wire. Roughly half of all copper mined is used to manufacture electrical wire and cable conductors.

- ➢ Many electrical devices rely on copper wiring because of its multitude of inherent beneficial properties, such as its high electrical conductivity, tensile strength, ductility, creep (deformation) resistance, corrosion resistance, low thermal expansion, high thermal conductivity, solderability, and ease of installation.

- ➢ Integrated circuits and printed circuit boards increasingly feature copper in place of aluminium because of its superior electrical conductivity.

- ➢ Heat sinks and heat exchangers use copper as a result of its superior heat dissipation capacity to aluminium.

- ➢ Electromagnets, vacuum tubes, cathode ray tubes, and magnetrons in microwave ovens use copper, as do wave guides for microwave radiation.

- ➢ Other uses of copper include: Utensils, Coin, Metal work, Refrigerator and Air Conditioning coils, Alloys e.g. bronze, brass.

- ➢ Copper proteins have diverse roles in biological electron transport and oxygen transportation, processes that exploit the easy interconversion of Cu(I) and Cu(II).

- ➢ Copper has also found medicinal use. It has been used from early times in the treatment of chest wounds and water purification. It has recently been suggested that copper helps to prevent inflammation associated with arthritis and such diseases.

- ➢ Copper is one of many trace elements required for good health. It is part of the prosthetic groups of many proteins and enzymes and thus is essential to their proper function. Since the body can not synthesize copper it must be taken in the diet. Nuts, seeds, cereals, meat (e.g. liver) and fish are good sources of copper.

- • **<u>Compounds of copper</u>**

Copper exhibits a variety of compounds, many of which are coloured. The two principal oxidation states of copper are +1 and +2 although some +3 complexes are known. Copper (I) compounds are expected to be diamagnetic in nature and are usually colourless, except where colour results from charge transfer or from the anion. The +1 ion has tetrahedral or square planar geometry. In solid compounds, copper (I) is often the more stable state at moderate temperatures.

The copper (II) ion is usually the more stable state in aqueous solutions. Compounds of this ion, often called cupric compounds, are usually coloured. They are affected by Jahn Teller distortions and exhibit a wide range of stereochemistries with four, five, and six coordination compounds predominating. The +2 ion often shows distorted tetrahedral geometry.

<u>**Oxides of copper**</u>

Copper (I) oxides are more stable than the copper (II) oxides at high temperatures. Copper (I) oxide occurs native as the red cuprite. In the laboratory, the reduction of Fehling's solution with a reducing sugar such as glucose produces a red precipitate. The test is sensitive enough for even 1 mg of sugar to produce the characteristic red colour of the compound. Cuprous oxide can also be prepared as a yellow powder by controlled reduction of an alkaline copper (II) salt with hydrazine. Thermal decomposition of copper (II) oxide also gives copper (I) oxide since the latter has greater thermal stability.

The same method can be used to prepare the compound from the copper (II) nitrate, carbonate and hydroxide. Copper (II) oxide occurs naturally as tenorite. This black crystalline solid can be obtained by the pyrolysis of the nitrate, hydroxide or carbonate salts. It is also formed when powdered copper is heated in air or oxygen. The table below shows some characteristics of copper oxides.

Copper oxides

Formula	Colour	Oxidation State	MP
CuO	black	Cu^{2+}	1026decomp
Cu_2O	red	Cu^{+}	1230

Copper halides

All of the copper (I) halides are known to exist although the fluoride has not yet been obtained in the pure state. The cuprous chlorides, bromides and iodides are coulouless, diamagnetic compounds. They crystallize at ordinary temperatures with the zinc blende structure in which Cu atoms are tetrahedrally bonded to four halogens. The copper (I) chloride and bromide salts are produced by boiling an acidic solution of copper (II) ions in an excess of copper. On dilution, the white CuCl or the pale yellow CuBr is

produced. Addition of soluble iodide to an aqueous solution of copper (II) ions results in the formation of a copper (I) iodide precipitate, which rapidly decomposes to Cu(I) and iodine. The copper (I) halides are sparingly soluble in water and much of the copper in aqueous solution is in the Cu (II) state. Even so, the poor solubility of the copper (I) compounds is increased upon addition of halide ions. The Table 11 below shows some properties of copper (I) halides.

Table 11. Melting and boiling points of Copper(II) halides

Formula	Colour	MP	BP	m (BM)	Structure
CuF_2	white	950decomp	-	1.5	
$CuCl_2$	brown	632	993decomp	1.75	$CdCl_2$
$CuBr_2$	black	498	-	1.3	

All four copper (II) halides are known although cupric iodide rapidly decomposes to cuprous iodide and iodine. The yellow copper (II) chloride and the almost black copper (II) bromide are the common halides. These compounds adopt a structure with infinite parallel bands of square CuX_4 units. Cupric chlorides and bromides are readily soluble in water and in donor solvents such as acetone, alcohol and pyridine.
Copper (II) halides are moderate oxidising agents due to the Cu (I)/ Cu (II) couple. In water, where the potential is largely that of the aqua-complexes, there is not a great deal of difference between them, but in non-aqueous media, the oxidising (halogenating) power increases in the sequence;
$CuF_2 \ll CuCl_2 \ll CuBr_2$.
They can be prepared by direct reaction with the respective halogens:

$Cu + F_2 \rightarrow CuF_2$;

$Cu + Cl_2 / 450\ C \rightarrow CuCl_2$;

$Cu + Br_2 \rightarrow CuBr_2$

Alternatively they can be prepared from CuX_2.aq by heating $\rightarrow CuX_2$

Table 12. Melting and boiling points of Copper(II) halides

Formula	Colour	MP	BP	Structure
CuCl	white	430	1359	-
CuBr	white	483	1345	-
CuI	white	588	1293	Zinc Blende

Prepared by reduction of $CuX_2^- > CuX$;   except for the F which has not been obtained pure. Note that CuI_2 has not been isolated because of the ease of reduction to CuI.

Copper (III) and (IV) compounds.

The best studied copper(III) compounds are the cuprate superconductors. Yttrium barium copper oxide ($YBa_2Cu_3O_7$) consists of both Cu(II) and Cu(III) centres. Like oxide, fluoride is a highly basic anion and is known to stabilize metal ions in high oxidation states. Indeed, both copper(III) and even copper(IV) fluorides are known, K_3CuF_6 and Cs_2CuF_6, respectively.

Coordination compounds of copper

Copper, like all metals, forms coordination complexes with ligands. In aqueous solution, copper(II) exists as $[Cu(H_2O)_6]^{2+}$. This complex exhibits the fastest water exchange rate (speed of water ligands attaching and detaching) for any transition metal aquo complex. Adding aqueous sodium hydroxide causes the precipitation of light blue solid copper(II) hydroxide. A simplified equation is:

$$Cu^{2+} + 2\ OH^- \rightarrow Cu(OH)_2$$

Aqueous ammonia results in the same precipitate. Upon adding excess ammonia, the precipitate dissolves, forming tetraamminecopper(II):

$$Cu(H_2O)_4(OH)_2 + 4\ NH_3 \rightarrow [Cu(H_2O)_2(NH_3)_4]^{2+} + 2\ H_2O + 2\ OH^-$$

Many other oxyanions form complexes; these include copper(II) acetate, copper(II) nitrate, and copper(II) carbonate. Copper(II) sulfate forms a blue crystalline pentahydrate, which is the most familiar copper compound in the laboratory.

<u>Organocopper compounds</u>

Compounds that contain a carbon-copper bond are known as organocopper compounds. They are very reactive towards oxygen to form copper(I) oxide and have many uses in chemistry. They are synthesized by treating copper(I) compounds with Grignard reagents, terminal alkynes or organolithium reagents.

<u>4.3.10. Zinc:</u>

Zinc has the symbol **Zn** (Fig. 46) and atomic number 30. It is the first element of group 12 of the periodic table. Zinc is, in some respects, chemically similar to magnesium, because its ion is of similar size and its only common oxidation state is +2. Brass, which is an alloy of copper and zinc, has been used since at least the 10th century BC. Zinc metal was not produced in large scale until the 12th century in India. Zinc is an essential mineral of "exceptional biologic and public health importance".

Fig. 46. Zinc metal

Zinc deficiency affects about two billion people in the developing world and is associated with many diseases.

- **<u>Occurrence of zinc</u>**

Fig. 47. Zinc sulphide ore

- The element is normally found in association with other base metals such as copper and lead in ores. Zinc is a chalcophile, meaning the element has a low affinity for oxides and prefers to bond with sulfides. Chalcophiles formed as the crust solidified under the reducing conditions of the early Earth's atmosphere. Sphalerite, which is a form of zinc sulfide (Figure shown in Fig. 47), is the most heavily mined zinc-containing ore because its concentrate contains 60–62% zinc.

Other minerals from which zinc is extracted include smithsonite (zinc carbonate), hemimorphite (zinc silicate), wurtzite (another zinc sulfide), and sometimes hydrozincite (basic zinc carbonate).

- **<u>Extraction of zinc</u>**

The Worldwide, 95% of the zinc is mined from sulfidic ore deposits, in which sphalerite ZnS is nearly always mixed with the sulfides of copper, lead and iron. Zinc metal is produced using extractive metallurgy. After grinding the ore, froth flotation, which selectively separates minerals from gangue by taking advantage of differences in their hydrophobicity, is used to get an ore concentrate. A final concentration of zinc of about 50% is reached by this process with the remainder of the concentrate being sulfur (32%), iron (13%), and SiO_2 (5%).

Roasting converts the zinc sulfide concentrate produced during processing to zinc oxide:

$$2\,ZnS + 3\,O_2 \rightarrow 2\,ZnO + 2\,SO_2$$

The sulfur dioxide is used for the production of sulfuric acid, which is necessary for the leaching process. If deposits of zinc carbonate, zinc silicate or zinc spinel, like the Skorpion Deposit in Namibia are used for zinc production the roasting can be omitted.

For further processing two basic methods are used: pyrometallurgy or electrowinning. Pyrometallurgy processing reduces zinc oxide with carbon or carbon monoxide at 950 °C (1,740 °F) into the metal, which is distilled as zinc vapor. The zinc vapor is collected in a condenser. The below set of equations demonstrate this process:

$$2\,ZnO + C \rightarrow 2\,Zn + CO_2$$
$$ZnO + CO \rightarrow Zn + CO_2$$

Electrowinning processing leaches zinc from the ore concentrate by sulfuric acid:

$$ZnO + H_2SO_4 \rightarrow ZnSO_4 + H_2O$$

After this step electrolysis is used to produce zinc metal.

$$2\,ZnSO_4 + 2\,H_2O \rightarrow 2\,Zn + 2\,H_2SO_4 + O_2$$

The sulfuric acid regenerated is recycled to the leaching step.

- ## **Properties of zinc**

Zinc, also referred to in nonscientific contexts as *spelter*, is a bluish-white, lustrous, diamagnetic metal, though most common commercial grades of the metal have a dull finish. It is somewhat less dense than iron and has a hexagonal crystal structure.

The metal is hard and brittle at most temperatures but becomes malleable between 100 and 150 °C. Above 210 °C, the metal becomes brittle again and can be pulverized by beating. Zinc is a fair conductor of electricity. For a metal, zinc has relatively low melting (419.5 °C, 787.1 F) and boiling points (907 °C). Its melting point is the lowest of all the transition metals aside from mercury and cadmium.

Zinc chemistry is similar to the chemistry of the late first-row transition metals nickel and copper, though it has a filled d-shell, so its compounds are diamagnetic and mostly colorless. The ionic radii of zinc and magnesium happen to be nearly identical. Because of this some of their salts have the same crystal structure and in circumstances where ionic radius is determining factor zinc and magnesium chemistries have much in common. Otherwise there is little similarity. Zinc tends to form bonds with a greater degree of covalency and it forms much more stable complexes with N- and S- donors. Complexes of zinc are mostly 4- or 6- coordinate although 5-coordinate complexes are known.

- ## **Applications of zinc**
  - Zinc is used as coating for the protection of Iron. Galvanization is used on chain-link fencing, guard rails, suspension bridges, lightposts, metal roofs, heat exchangers, and car bodies.

➢ The relative reactivity of zinc and its ability to attract oxidation to itself makes it an efficient sacrificial anode in cathodic protection (CP). For example, cathodic protection of a buried pipeline can be achieved by connecting anodes made from zinc to the pipe.

➢ Zinc is used as an anode material for batteries.

➢ Powdered zinc is used in this way in alkaline batteries and sheets of zinc metal form the cases for and act as anodes in zinc–carbon batteries.

➢ Zinc is used as the anode or fuel of the zinc-air battery/fuel cell.

➢ A widely used alloy which contains zinc is brass. It is useful in communication equipment, hardware, musical instruments, and water valves.

➢ Zinc oxide is widely used as a white pigment in paints, and as a catalyst in the manufacture of rubber.

➢ Zinc sulfide (ZnS) is used in luminescent pigments such as on the hands of clocks, X-ray and television screens, and luminous paints.

➢ In humans, zinc plays "ubiquitous biological roles". It interacts with "a wide range of organic ligands", and has roles in the metabolism of RNA and DNA, signal transduction, and gene expression.

- ## **Compounds of zinc**

Zinc (I) compounds are rare, and requires bulky ligands to stabilize the low oxidation state. Most zinc (I) compounds contains formally the $[Zn_2]^{2+}$ core, which is analogous to the $[Hg_2]^{2+}$ dimeric cation present in mercury (I) compounds. The diamagnetic nature of the ion confirms its dimeric structure. The first zinc (I) compound containing the Zn—Zn bond, $(\eta^5\text{-}C_5Me_5)_2Zn_2$, is also the first dimetallocene. The $[Zn_2]^{2+}$ ion rapidly disproportionates into zinc metal and zinc(II), and has only been obtained as a yellow glass formed by cooling a solution of metallic zinc in molten $ZnCl_2$.

Binary compounds of zinc are known for most of the metalloids and all the nonmetals except the noble gases. The oxide ZnO is a white powder that is nearly insoluble in neutral aqueous solutions, but is amphoteric, dissolving in both strong basic and acidic solutions. The other chalcogenides (ZnS, ZnSe, and ZnTe) have varied applications in electronics and optics. Pnictogenides (Zn_3N_2, Zn_3P_2, Zn_3As_2 and Zn_3Sb_2), the peroxide (ZnO_2), the hydride (ZnH_2), and the carbide (ZnC_2) are also known. Of the four halides, ZnF_2 has the most ionic character, whereas the others ($ZnCl_2$, $ZnBr_2$, and ZnI_2) have relatively low melting points and are considered to have more covalent character.

In weak basic solutions containing Zn^{2+} ions, the hydroxide $Zn(OH)_2$ forms as a white precipitate. In stronger alkaline solutions, this hydroxide is dissolved to form zincates ($[Zn(OH)_4]^{2-}$). The nitrate $Zn(NO_3)_2$, chlorate $Zn(ClO_3)_2$, sulfate $ZnSO_4$, phosphate $Zn_3(PO_4)_2$, molybdate

$ZnMoO_4$, cyanide $Zn(CN)_2$, arsenite $Zn(AsO_2)_2$, arsenate $Zn(AsO_4)_2 \cdot 8H_2O$ and the chromate $ZnCrO_4$ (one of the few colored zinc compounds) are a few examples of other common inorganic compounds of zinc. One of the simplest examples of an organic compound of zinc is the acetate.

4.4 EXERCISE QUESTIONS

Exercise Questions: Type 1

1. Write the electronic configuration of titanium in titanium nitride.
2. Give the formula of the oxide of osmium in its highest oxidation state.
3. What is responsible for the color of aqueous solutions of the transition metal ions?
4. What is the value of spin only magnetic moment for $[CoF_6]^{3-}$ complex ion.
5. Name the transition metal oxide catalyst used in the contact process of manufacture of sulphuric acid.
6. Which is the structure obtained by the dsp^2 hybridization of central metal atom/ion in a complex?
7. Mention one ore of cobalt with chemical formula
8. What is the role of coke in the extraction of iron from its ore?
9. Give the chemical formula of Prussian blue.
10. What is the role of silica in the extraction of copper from copper pyrites?
11. Name the transition metal catalyst used for the hydrogenation process.
12. Zinc salts are colourless. Give reason.
13. Define transition elements?
14. Give the electronic configurations of Cr^{3+} and Co^{2+}.
15. The densities of transition elements are high. Justify.
16. All the transition elements have high melting and boiling points as compared to s-block elements. Give reason.
17. Give the electronic configurations of Nb and Ag.
18. Transition elements form variable oxidation states. Justify.
19. What is responsible for the colour of the transition metal complexes?
20. Potassium ferrocyanide exhibits diamagnetic nature. Give reason.
21. Which pair of catalyst is used for polymerization reaction?
22. What is the hybridization involved in $[Cu(NH_3)_4]SO_4$ complex?
23. What is the structure obtained by d^2sp^3 hybridisation?
24. Mention any two applications of non-stoichiometric compounds.
25. Give the composition of perovskite ore.
26. Give an example for the compound of chromium which is amphoteric in nature.
27. Give the composition of important ore of manganese.

Exercise Questions: Type 2

1. Explain the variation of atomic radii of 1st row transition series
2. Group oxidation number is achieved by the elements that lie towards the left of the d block but not by the elements on the right. Give reason.
3. What do you mean by inner orbital complex? Give an example for outer orbital complex.
4. Name three important factors that affect the magnitude of crystal field splitting energy (Δo) in coordination complexes.
5. Explain the extraction of titanium by Kroll process.
6. Mention any four applications of iron.
7. Explain the Mond's process of extraction of nickel from the nickel ores.
8. Write a short note on copper (II) halides.
9. Explain how zinc is extracted from its ore by pyrometallrgical process.
10. What are transition metal halide clusters? Give one example.

11. $[NiCl_4]^{2-}$ is high spin complex, while $[NiCN_4]^{2-}$ is low spin complex. Justify.

Exercise Questions: Type 3

1. Explain briefly the property of variable oxidation states of 3d-transition metals.
2. Explain very briefly about the extraction of chromium from chromite and manganese from pyrolusite ores respectively.
3. Mention any one important ore of scandium and vanadium with their chemical composition.
4. Write a short note on applications of chromium.
5. Explain the chemistry involved in the extraction of iron using the blast furnace, with reactions.
6. Explain the applications of the following metals in various sectors,
    i. Copper
    ii. Zinc
7. Explain the structure and bonding involved in $K_2Re_2Cl_8$.
8. Describe what happens when the following species are acidified and then treated with zinc and account for all your observations:
    i. ammonium vanadate (V)
    ii. potassium chromate (VI)
9. Describe what happens when the following species are treated with sodium hydroxide and hydrogen peroxide and account for all your observations:
    i. cobalt (II) chloride
    ii. chromium (III) chloride
10. Describe what happens when cobalt chloride is treated with concentrated ammonia solution and then left to stand in air. Account for all your observations.
11. Describe two ways in which the concentration of Fe^{2+} in a sample could be determined by titration. In each case write an equation for the reaction occurring during the titration.
12. Write a note on variable oxidation states exhibited by transition metals.

13. Explain the structure and bonding in $Ni(CO)_4$ complex using Valence bond theory. Explain the magnetic properties of the following complexes:
$K_2[MnCl_4]$, $[Fe(CN)6]^{3-}$ and $[FeF_6]^{3-}$.

Chapter **5**

INNER TRANSITION ELEMENTS (*f*- BLOCK ELEMENTS)

5.1 INTRODUCTION TO LANTAHNIDES

Lanthanides (also called lanthanoids) are called first inner transition series or third transition and come immediately after lanthanum. Inner transition elements are those elements in which the last electron enters the f- orbital. The elements in which 4f (lanthanides) and 5f (actinides) orbitals are progressively filled are called f-block elements. Lanthanides are classified as f block elements along with the actinides. They are commonly called the rare earths. They are characterized by the filling up of the 4f energy levels which are not usually involved in bonding. These highly electropositive elements have a common oxidation state of +3 and generally resemble each other in their chemical and physical properties. They have a generic symbol "Ln". The general electronic configuration of lanthanides is [Xe] $4f^{0-14}$ $5d^{0-1}$ $6s^2$. The general electronic configuration of lanthanides can also be written as [Xe] $(n-2)f^{0-14}$ $(n-1)d^{0-1}$ ns^2.

Discovery and Occurrence

Atomic Number	Name	Symbol
57*	lanthanum*	La
58	cerium	Ce
59	praseodymium	Pr
60	neodymium	Nd
61	promethium	Pm
62	samarium	Sm
63	europium	Eu
64	gadolinium	Gd
65	terbium	Tb
66	dysprosium	Dy
67	holmium	Ho
68	erbium	Er
69	thulium	Tm
70	ytterbium	Yb
71	lutetium	Lu

In 1794, Swedish chemist, Gadolin, discovered an oxide called yttria. Yttria was broken down to– yttria, erbis and terbia. Over the years, more sparation was achieved, and more oxide discovered e.g. lutetia. The names were later obtained by changing the ending –a to -um. Major sources of lanthanides are Monazite sand-composed of phosphates of thorium, cerium, neodymium and lanthanum; the phosphate portion of monazite contains small traces of other lanthanide ions and the only lanthanide that does not occur naturally is promethium, which is made artificially by nuclear reaction. Bastnaesite, found in USA and Madagascar is a mixed fluorocarbonate $M_{III}CO_3F$ where M is La or the lanthanide metals. It provides 20% total supply of lanthanides. Also very small amount of xenotime mineral is mined.

5.2 EXTRACTION AND SEPARATION OF LANTHANIDES

They are extracted from the earlier mentioned ores. Monazite is treated with hot concentrated H_2SO_4. Th, La and the Ln dissolve as sulphates and are separated from insoluble material. Th is precipitated as ThO_2 by partial neutralization with NH_4OH. Na_2SO_4 is used to salt out La and the lighter Ln as sulphates, leaving the heavy lanthanides in solution. The light Ln are oxidized with bleaching powder $Ca(OCl)_2$. Ce^{2+} is oxidized to Ce^{4+} which is precipitated as $Ce(IO_3)_4$ and removed. The extraction process from bastnaesite is slightly simpler since it does not contain Th. The different lanthanides elements can be separated by various methods;

Reduction of their Trihalides: La, Ce, Pr, Nd and Gd may be obtained by reduction of their trichlorides with calcium at about 1000°C in an argon filled vessel e.g.

$$2MCl_3 + 3Ca \rightarrow 2M + 3CaCl_2 \text{ (T > 1000°C)}$$

The heavier Ln like Tb, Dy, Ho, Er and Tm can also be obtained by this method but the trifluorides is used, since their trichloride is volatile.

$$2MF_3 + 3Ca \rightarrow 2M + 3CaF_2 \text{ (}MCl_3 \text{ is too volatile)}$$

Eu, Sm and Yb are obtained by chemical reduction of their trioxides.

$$M_2O_3 + 2La \rightarrow 2M + La_2O_3 \text{ (}MCl_3 \text{ reduced to } MCl_2 \text{ by Ca)}$$

Ion exchange: The basis of the lanthanide series separation on an ion exchange column is their ability to form complex ions. All lanthanides form +3 ions, M^{+3} whose ionic radii decrease progressively with increasing atomic number from Ce^{+3} to Lu^{+3}. As a solution containing +3 lanthanides ions is placed at the top of a column of cation exchange resin [e.g. is Dowex-50 made of a sulphonated polystyrene and contains functional groups $-SO_3H$.] The Ln^{+3} ions are absorbed into the resin and an equivalent amount of hydrogen ions are released from the column;

$$Ln^{+3}(aq) + 3H^+ R^-(s) \rightarrow Ln(R^-)(s) + 3H^+(aq)$$

A citrate buffer (citric acid/ammonium citrate) solution (which complexes with the lanthanide ions) is slowly run down the column and the cations partition themselves between the column itself and the moving citrate solution. Since the smaller ions show a greater preference for complexing with the citrate solution, these ions are the first to emerge from the column. By the correct choice of conditions the lutetium ion, Lu^{+3} (aq), emerges first from the column, followed by the cations ytterbium, thulium, erbium, etc, in order of increasing ionic radius. By using a long ion-exchange column, the elements may be obtained at 99.9% with one pass.

Valency change: The different properties of the various oxidation states make separation very easy [i.e, the properties of Ln^{+4} and Ln^{+2} are very different from that of Ln^{+3}]. Cerium can be separated from Ln mixtures because it is the only one which has a Ln^{+4} ions stable in aqueous solution. A solution containing mixture of Ln^{+3} ions can be oxidized with NaOCl under alkaline conditions to produce Ce^{+4}. Because of the higher charge, Ce^{+4} is much smaller and less basic than Ce^{+3} or any other Ln^{+3}. The Ce^{+4} is separated by carefully controlled precipitation of CeO_2 or $Ce(IO_3)_4$, leaving the trivalent ions in solution. Also, Eu^{2+} can be separated from a mixture of Ln^{+3}. If a solution of Ln^{+3} ions Cation is reduced electrolytically using a Hg cathode or Zn amalgam, then Eu_{2+} is produced. If H_2SO_4 is present, $EuSO_4$ which is insoluble will be precipitate. This can be filtered off. Other methods are Solvent Extraction, Precipitation, Thermal reaction, Fractional crystallization, Complex formation.

5.3 GENERAL CHARACTERISTICS OF LANTAHNIDES

Lanthanide contraction

Each succeeding lanthanides differs from its immediate predecessor in having one or more electron in the 4f (though there are some exceptions) and an extra proton in the nucleus of the atom. The 4f electrons constitute inner shells and are rather ineffective in screening the nucleus. This leads to a gradual increase in the

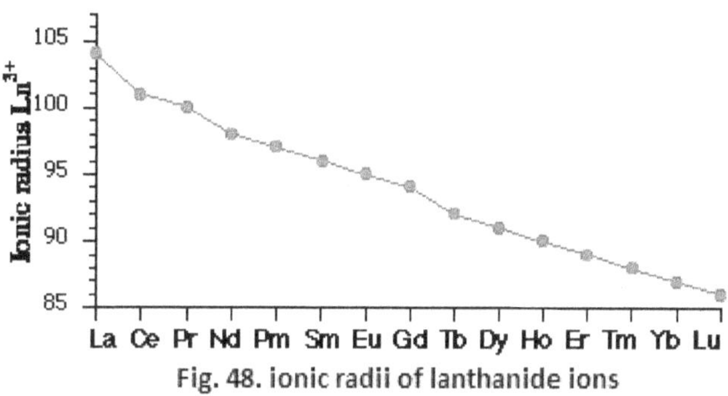

Fig. 48. ionic radii of lanthanide ions

attraction of the nucleus to the electrons in the outermost shell as the nuclear charge increases, and a consequent contraction in the atomic radius. This steady decrease in the atomic size of lanthanides with the increase in the atomic number along the 4f series from cerium to lutetium is called lanthanide contraction. As a consequence, even the ionic size of Ln^{3+} ions also gradually decreases as shown in the above picture.

Consequences of lanthanide contractions

1. *Similarity in the properties of elements of second and third transition series.* Similarity in properties, with gradual changes occurring across the lanthanide series: a size effect from the Lanthanide Contraction.
2. *Separation of lanthanides is possible due to lanthanide contraction.* As the ionic radii contract along the lanthanide series, the ability to form complex ions increases and this is the basis of their separation on an ion exchange column.
3. *Variation in the basic strength of lanthanide hydroxides.* It is due to lanthanide contraction that there is variation in the basic strength of lanthanide hydroxides. The basic strength decreases from $La(OH)_3$ to $Lu(OH)_3$.

Properties of the lanthanide metals

Lanthanide metals are silvery white, but tarnish in air. Rather soft (later M are harder). They have high melting & boiling points. They are Very reactive: $(I_1 + I_2 + I_3)$ comparatively low. They burn easily in air, but slowly in cold. They burn Burn at $T > 150\ °C$. Their reaction with H_2 is exothermic reaction. They react readily with C, N_2, Si, P, halogens & other non-metals. They form binaries on heating with most non-metals (e.g. LnN, Ln_2S_3, LnB_6, LnC_2,..).

The chemical properties of these elements are governed by their valence electrons (in their outermost shells). Each rare earth element has two s-electrons in its outermost shell and either eight or nine electrons in the next shell inward (two s, six p, and zero or one d-electron). This accounts for the chemical similarity of these elements. Most form M^{+3} ions by loss of the two outer s-electrons and either the d-electron one shell inward or one of the f-electrons two shells inwards.

❖ *Physical Properties*

1. The metals are silvery white in colour.
2. They generally have high melting and boiling points and are very hard
3. They are good conductors of heat and electricity.
4. Many of the lanthanide ions form coloured ions.
5. The lanthanides exhibit a principal oxidation state of +3 in which the M^{+3} ion contains an outer shell containing 8 electrons and an underlying layer containing up to 14- 4f electrons.
6. They exhibit paramagnetism because of the presence of unpaired electrons.

❖ *Chemical Properties*

In chemical reactivity, they resemble calcium.
1. They readily tarnish in air and burn to give oxides (all give trioxides except Ce which forms CeO_2).
2. They also combine with the following non-metals –N, S, halogens, H.
3. The hydrides are non-stoichiometric but have a composition of MH_3. These hydrides liberate hydrogen from water.
4. The lanthanides also liberate hydrogen from water as does their hydrides and a vigorous evolution of same gas from dilute non-oxidizing acids.
5. Lanthanide compounds are generally predominantly ionic and usually contain lanthanide metal in its +3 oxidation states.

Electronic structure

La has one 5d and two 6s electrons; the next element cerium, however, while still retaining two 6s electrons has two electrons in the 4f orbital and none in the 5d orbital. The atoms of the elements from cerium to lutetium have between 2 and 14 electrons in the 4f orbital. The filling up of the 4f orbital is regular with one exception; Europium has the outer electronic structure of $[Xe]4f^7 5d^0 6s^2$ and the next element gadolinium has an extra electron in the 5d shell, $[Xe]4f^7 5d^1 6s^2$ [when all seven 4f shells are singly occupied, a degree of stability is conferred on the atom]. Ytterbium has a full complement of 4f electrons ($[Xe] 4f^{14} 5d^0 6s^2$) and the extra electron in the lutetium atom enters the 5d shell ($[Xe] 4f^{14} 5d^1 6s^2$). Only lanthanum, gadolinium and lutetium have single 5d electrons. The other lanthanides do not have electrons on their 5d orbitals.

<table>
<tr><td colspan="6" align="center">Table 13. Propeties of lanthanides</td></tr>
<tr><td>Atomic Number</td><td>Name</td><td>Symbol</td><td>Electron Configuration</td><td>Formula of Chloride</td><td>Ox. State of Lanthanide</td></tr>
<tr><td>57*</td><td>lanthanum*</td><td>La</td><td>$[Xe]\ 5d^1\ 6s^2$</td><td>$LaCl_3$</td><td>+3</td></tr>
<tr><td>58</td><td>cerium</td><td>Ce</td><td>$[Xe]\ 4f^2\ 5d^0\ 6s^2$</td><td>$CeCl_3,\ CeCl_4$</td><td>+3, +4</td></tr>
<tr><td>59</td><td>praseodymium</td><td>Pr</td><td>$[Xe]\ 4f^3\ 5d^0\ 6s^2$</td><td>$PrCl_3,\ PrCl_4$</td><td>+3, +4</td></tr>
<tr><td>60</td><td>neodymium</td><td>Nd</td><td>$[Xe]\ 4f^4\ 5d^0\ 6s^2$</td><td>$NdCl_3$</td><td>+3</td></tr>
<tr><td>61</td><td>promethium</td><td>Pm</td><td>$[Xe]\ 4f^5\ 5d^0\ 6s^2$</td><td>$PmCl_3$</td><td>+3</td></tr>
<tr><td>62</td><td>samarium</td><td>Sm</td><td>$[Xe]\ 4f^6\ 5d^0\ 6s^2$</td><td>$SmCl_2,\ SmCl_3$</td><td>+2, +3</td></tr>
<tr><td>63</td><td>europium</td><td>Eu</td><td>$[Xe]\ 4f^7\ 5d^0\ 6s^2$</td><td>$EuCl_2,\ EuCl_3$</td><td>+2, +3</td></tr>
<tr><td>64</td><td>gadolinium</td><td>Gd</td><td>$[Xe]\ 4f^7\ 5d^1\ 6s^2$</td><td>$GdCl_3$</td><td>+3</td></tr>
<tr><td>65</td><td>terbium</td><td>Tb</td><td>$[Xe]\ 4f^9\ 5d^0\ 6s^2$</td><td>$TbCl_3,\ TbCl_4$</td><td>+3, +4</td></tr>
<tr><td>66</td><td>dysprosium</td><td>Dy</td><td>$[Xe]\ 4f^{10}\ 5d^0\ 6s^2$</td><td>$DyCl_3$</td><td>+3</td></tr>
<tr><td>67</td><td>holmium</td><td>Ho</td><td>$[Xe]\ 4f^{11}\ 5d^0\ 6s^2$</td><td>$HoCl_3$</td><td>+3</td></tr>
<tr><td>68</td><td>erbium</td><td>Er</td><td>$[Xe]\ 4f^{12}\ 5d^0\ 6s^2$</td><td>$ErCl_3$</td><td>+3</td></tr>
<tr><td>69</td><td>thulium</td><td>Tm</td><td>$[Xe]\ 4f^{13}\ 5d^0\ 6s^2$</td><td>$TmCl_3$</td><td>+3</td></tr>
<tr><td>70</td><td>ytterbium</td><td>Yb</td><td>$[Xe]\ 4f^{14}\ 5d^0\ 6s^2$</td><td>$YbCl_2,\ YbCl_3$</td><td>+2, +3</td></tr>
<tr><td>71</td><td>lutetium</td><td>Lu</td><td>$[Xe]\ 4f^{14}\ 5d^1\ 6s^2$</td><td>$LuCl_3$</td><td>+3</td></tr>
</table>

Oxidation States: Lanthanides exhibit a principal oxidation state of +3 which contain an outer shell containing 8 electrons and an underlying layer containing up to 14 electrons. The +3 ions of La, Gd and Lu which contain respectively an empty, a half-filled, and a completely filled 4f level are especially stable. Ce can exhibit an oxidation state of +4 in which it has the same electronic structure with La^{+3} i.e. an empty 4f level-noble gas configuration). Also, Tb^{4+} exists which has the same electronic structure as Gd^{3+} i.e. a half-filled 4f level. An empty, a half filled and a completely filled 4f shell confers some extra stability on a particular oxidation state.

La^{3+} $\quad$ $4f^0 5d^0 6s^0$ $\qquad$ Ce^{4+} $\quad$ $4f^0 5d^0 6s^0$ $\quad$ (empty 4f level)

Gd^{3+} $\quad$ $4f^7 5d^0 6s^0$ $\qquad$ Tb^{4+} $\quad$ $4f^7 5d^0 6s^0$ $\quad$ (half-filled 4f level)

Also, Eu^{+2} is iso-electronic with Gd^{+3} i.e. half-filled 4f level and Yb^{+2} is iso-electronic with Lu^{+3}

Gd^{3+} $\quad$ $4f^7 5d^0 6s^0$ $\qquad\qquad$ Eu^{2+} $\quad$ $4f^7 5d^0 6s^0$ $\quad$ (half-filled 4f level)

Lu^{3+} $\quad$ $4f^{14} 5d^0 6s^0$ $\qquad$ Yb^{2+} $\quad$ $4f^{14} 5d^0 6s^0$ $\quad$ (completely filled 4f level)

In addition, +2 and +4 states exist for elements that are close to these states. For example, Sm^{2+} and Tm^{2+} occur with f^6 and f^{13} arrangements and Pr^{4+} and Nd^{4+} have f^1 and f^2 arrangements. The most stable oxidation state is Ln^{3+} and Ln^{2+} and Ln^{4+} are less stable. Ce^{4+} is strongly oxidizing and Sm^{2+} is strongly reducing:

$$Ce^{4+} + Fe^{2+} \rightarrow Ce^{3+} + Fe^{3+}$$

]

$$2Sm^{2+} + 2H_2O \rightarrow 2Sm^{3+} + 2OH^- + H_2$$

{Ce^{4+} and Sm^{2+} are converted to +3 state, showing that it is the most stable oxidation state}

Colour of salts

Salts of many lanthanide elements are coloured. These colours are seen to depend on the electronic arrangement in the *4f* orbital. When this orbital is empty, half-filled or full, such ions are found to be colourless. Thus La^{3+} ion (f^0), Gd^{3+} (f^7) and Lu^{3+} (f^{14}) are colourless.

In the case of other ions, it is seen that configurations with f^n and f^{14-n} have the same colour, because they have the same number of unpaired electrons. Since the f-orbitals are well protected by the 5th and 6th electronic shells, they are split by ligand fields only to a lesser extent compared to the d-orbital splitting in the outer transition elements. The colours are due to low-energy $f - f$ electronic transitions.

Table 14. Colours of lanthanide ions		
Ions	configuration	colour
La^{3+}	f^0	Colour less
Ce^{3+}	f^1	Colour less
Pr^{3+}	f^2	Green
Nd^{3+}	f^3	Red
Pm^{3+}	f^4	Pink
Sm^{3+}	f^5	Yellow
Eu^{3+}	f^6	Colour less
Gd^{3+}	f^7	Colour less
Tb^{3+}	f^8	Colour less
Dy^{3+}	f^9	Yellow
Ho^{3+}	f^{10}	Pink
Er^{3+}	f^{11}	Red
Tm^{3+}	f^{12}	Green
Yb^{3+}	f^{13}	Colour less
Lu^{3+}	f^{14}	Colour less

<u>**Magnetic properties**</u>

In addition to the spin, motion of electrons in the orbital around the nucleus also generates a magnetic moment. The portion of the total magnetic moment generated by the electron spin is called the *"spin contribution to magnetic moment"* and the portion due to the motion around the nucleus is called the *"orbital contribution to magnetic moment."* The total magnetic moment of the atom or ion will be the total *sum of these two contributions for all the unpaired electrons* in it. For a single electron, its spin contribution depends on its spin quantum number, and its orbital contribution depends on the magnetic quantum numbers available in its orbital. The possible electronic transitions generating the spectrum of the atom or ion also depend on these quantum numbers. If s is the spin quantum number of an electron, its spin contribution to magnetic moment, μs, is given by: $\mu_s = g\sqrt{s(s+1)}$

where s is the absolute value of the spin (i.e. ½) and 'g' is a constant called the *"gyromagnetic ratio."* The orbital contribution to the magnetic moment for an electron, μl, depends on the l value of its orbital, and is given by: $\mu_l = \sqrt{l(l+1)}$

These two moments will couple with each other to produce the total magnetic moment due to this electron. The total effect can then be represented by a new quantum number 'j', where $j = (l + s)$. Magnetic moment of a J-state is expressed by the Landé formula:

$$\mu = g_J\sqrt{J(J+1)}\,\mu_B \quad g_J = \frac{3}{2} + \frac{S(S+1)-L(L+1)}{2J(J+1)}$$

The 4f electrons are responsible for the strong magnetism exhibited by the metals and compounds of the lanthanides. In the incomplete 4f subshell the magnetic effects of the different electrons do not cancel out each other as they do in a completed subshell, and this factor gives rise to the interesting magnetic behaviour of these elements. At higher temperatures, all the lanthanides except lutetium are paramagnetic (weakly magnetic), and this paramagnetism frequently shows a strong anisotropy. As the temperature is lowered, many of the metals exhibit a point below which they become antiferromagnetism (i.e., magnetic moments of the ions are aligned but some are opposed to others), and, as the temperatures are lowered still further, many of them go through a series of spin rearrangements, which may or may not be in conformity with the regular crystal lattice. Finally, at still lower temperatures, a number of these elements become ferromagnetic (i.e., strongly magnetic, like iron).

Table 15. Magnetic moments of lanthanie ions

Ions	configuration	No. of unpaired electrons	Observed μ_B
La^{3+}	f^0	0	0
Ce^{3+}	f^1	1	2.3-2.5
Pr^{3+}	f^2	2	3.4-3.6
Nd^{3+}	f^3	3	3.5-3.6
Pm^{3+}	f^4	4	-
Sm^{3+}	f^5	5	1.4-1.7
Eu^{3+}	f^6	6	3.3-3.5
Gd^{3+}	f^7	7	7.9-8.0
Tb^{3+}	f^8	6	9.5-9.8
Dy^{3+}	f^9	5	10.4-10.6
Ho^{3+}	f^{10}	4	10.4-10.7
Er^{3+}	f^{11}	3	9.4-9.6
Tm^{3+}	f^{12}	2	7.1-7.6
Yb^{3+}	f^{13}	1	4.3-4.9
Lu^{3+}	f^{14}	0	0

Applications of lanthanides

1. **Use as catalysts**: Lanthanide catalysts have been repeatedly recommended for use in numerous organic reactions, including the hydrogenation of ketones to form secondary alcohols, the hydrogenation of olefins to form alkanes, the dehydrogenation of alcohols and butanes, and the formation of polyesters.

2. **Use in the glasses industry**: Cerium oxides has been found to be a more rapid polishing agent for glass & are used in the polishing of lenses for cameras, binoculars, and eyeglasses, as well as in polishing mirrors and television faceplates. Praseodymium and neodymium are added to glass to make welders' and glassblowers' goggles, that absorb the bright-yellow light from the sodium flame. The same combination is sometimes added to the glass used in television faceplates to decrease the glare from outside light sources. Cerium oxide increases the opacity of white porcelain enamels.

3. **Use in the metallurgical industry**: Small amounts of misch metal and cerium have long been added to other metals or alloys to remove their nonmetallic impurities. Misch metal added to cast iron makes a more malleable nodular iron. Added to some steels, it makes them less brittle. The addition of misch metal to certain alloys has been reported to increase the tensile strength and improve the hot workability and the high-temperature oxidation resistance.

4. **Use in the television industry**: It has been found that if a small amount of europium oxide (Eu_2O_3) is added to yttrium oxide (Y_2O_3), it gives a brilliant-red phosphor. Colour television screens utilize red, green, and blue phosphors. The Y_2O_3–Eu_2O_3 phosphor corrected these disadvantages and made possible much brighter and more natural coloured pictures. The rare-earth phosphors are also finding use in mercury-arc lights, which are used for sporting events and special street lighting. Considerable amounts of mixed rare-earth fluorides are used to make cored carbon rods, which are used as arcs in searchlights and in some of the lights used by the motion-picture industry.

5. **Use as NMR shift reagents**: Paramagnetism of lanthanide ions is utilized to spread resonances in H^1 NMR of organic molecules that coordinate to lanthanides.

6. **Other applications**: Another significant industrial application of rare earths is in the manufacture of strong permanent magnets. Alloys of cobalt with rare earths, such as cobalt– samarium, produce permanent magnets that are far superior to most of the varieties now on the market. Another relatively recent development is the use of a barium phosphate– europium phosphor in a sensitive X-ray film that forms satisfactory images with only half the exposure. Europium, gadolinium, and dysprosium have large capture cross sections for thermal neutrons. Complexes of europium, praseodymium, or ytterbium with derivatives of camphor are useful reagents for analysis of optically-active organic compounds, which often are obtained as mixtures containing unknown proportions of two components that differ only in that their molecular structures are mirror images of each other.

5.4 INTRODUCTION TO ACTINIDES

Actinides (also called actinoids) are called second inner transition series or fourth transition and come immediately after actinium. This group is also known as heavier elements or inner series transition elements following actinium element.

The actinides are all radioactive. Uranium (Z = 92) is the upper limit of naturally occurring elements. Of the 14 actinides, only U and Th are found in appreciable quantities in nature. All elements of higher atomic mass (the 'transuranium' elements) have only been synthesized in small amounts in particle accelerators. Macroscopic samples of mendelevium (Md), nobelium (No) and lawrencium (Lr) have never been seen. Actinides are silvery and chemically reactive. Like the lanthanides, they form highly colored compounds.

5.5 SYNTHESIS AND STABILITY OF HEAVIER ELEMENTS

Table 16. Actinide elements		
Atomic Number	Name	Symbol
89	Actinium	Ac
90	Thorium	Th
91	Protactinium	Pa
92	Uranium	U
93	Neptunium	Np
94	Plutonium	Pu
95	Americium	Am
96	Curium	Cm
97	Berkelium	Bk
98	Californium	Cf
99	Einsteinium	Es
100	Fermium	Fm
101	Mendelevium	Md
102	Nobelium	No
103	Rutherfordium	Rf

In 1934-35, Enrico Fermi and coworkers observed that when most elements are bombarded with slow neutrons, they are converted into radioactive products which emit β-rays on decay. Since a β-emission leads to an increase of one unit in the nuclear positive charge, they used uranium as the target for neutrons and produced elements with atomic numbers greater than 92 for the first time. Before this time, only three members of the actinide series were known and the existence of such a series after the lanthanides was not suspected. In 1945, it was G. T. Seaborg of the University of California who suggested the probable existence of such a series of elements. However, the elements within this group show a much wider variation in oxidation states, and therefore greater differences in chemical properties, than the elements in the lanthanide series.

Although more than ten transuranium elements have been discovered, only the first six have been studied extensively for their chemical properties, and only one, plutonium, has been prepared in quantity. The others have been obtained only in very small quantities and studied using tracers or ultra microchemical techniques. The dangerous level of α-activity in these isotopes makes them unsafe for handling without special equipment.

Occurrence and Preparation of actinides: With the exception of Th and U (natural) the actinoids are man made, synthesized by artificial radioactivity i.e by bombardment with slow moving neutron. Only Actinium, Thorium, Protactinium & Uranium occur naturally (i.e. Z ≤ 92).

Neptunium (93): Prepared in 1940, McMillan and Abelson at the University of California by bombarding a layer of uranium oxide with slow neutrons from a cyclotron.

$$_{92}U^{238} + _0n^1 \rightarrow _{92}U^{239} \qquad \text{(half life 23 min.)}$$

$$_{92}U^{239} \rightarrow _{93}Np^{239} + _{-1}e^0 \qquad \text{(half life 2.3 days.)}$$

A more stable isotope of neptunium was prepared in weighable quantities by Seaborg and Wahl in 1942 by the action of high-speed neutrons on uranium.

$$_{92}U^{238} + _0n^1 \rightarrow _{92}U^{237} + 2\,_0n^1 \qquad \text{(half life 7 days.)}$$

$$_{92}U^{237} \rightarrow _{93}Np^{237} + _{-1}e^0 \qquad \text{(half life 2.25 x }10^6 \text{ years.)}$$

Plutonium (94): Seaborg, McMillan, Wahl and Kennedy synthesized this element in 1940 by bombarding uranium with deuterium.

$$_{92}U^{238} + _1H^2 \rightarrow _{93}Np^{238} + 2\,_0n^1 \qquad \text{(half life 2 days.)}$$

$$_{93}Np^{238} \rightarrow _{94}Pu^{238} + _{-1}e^0 \qquad \text{(half life 90 years.)}$$

But another isotope of plutonium with a longer life is produced in large quantities inside nuclear reactors using enriched uranium as fuel. Although the fissionable material in the fuel is U235, some of the neutrons produced are absorbed by the U238 present, which then changes into plutonium:

$$_{92}U^{238} + _0n^1 \rightarrow _{92}U^{239} \qquad \text{(half life 23 min.)}$$

$$_{92}U_{239} \rightarrow _{93}Np^{239} + _{-1}e^0 \qquad \text{(half life 2.3 days.)}$$

$$_{93}Np^{239} \rightarrow _{94}Pu^{239} + _{-1}e^0 \qquad \text{(half life 2.4 x }10^4 \text{ years.)}$$

The plutonium produced is fissionable and is a much better nuclear fuel than U^{235}. Such reactors which use U^{235} fuel to produce plutonium are therefore called "breeder reactors." The fuel used in the atom bombs dropped in Hiroshima and Nagasaki was plutonium. It is a strong α-emitter producing about 140 million disintegrations per minute per milligram of material, which can be destructive to living tissues.

Americium (95): Seaborg, James and Morgan synthesized this element in 1944 by bombarding U^{238} with high-speed (40 Mev) α-particles from a cyclotron.

$$_{92}U^{238} + _2He^4 \rightarrow _{94}Pu^{241} + _0n^1 \qquad \text{(half life 13 years.)}$$

$$_{94}Pu_{241} \rightarrow _{95}Am^{241} + _{-1}e^0 \qquad \text{(half life 470 years.)}$$

The α-activity of this species is approximately 70 billion disintegrations per minute per milligram of material. A less dangerous and more stable isotope Am^{243} has been produced with a half life of about 10,000 years.

Curium (96): This was identified by Seaborg, James and Ghiorso in 1944 in products obtained by bombarding plutonium with α-particles from the Berkeley cyclotron.

$$_{94}Pu^{239} + {_2}He^4 \rightarrow {_{96}}Cm^{242} + {_0}n^1 \qquad \text{(half life 162 days.)}$$

The α-activity is approximately 1014 disintegrations per minute per milligram, which is not only dangerous to tissues, but also affects the solvents used to study its chemistry. A more stable isotope Cm^{243} with a half life of 100 years has also been reported.

Berkelium (97): This element was discovered in 1949 by Seaborg, Thompson and Ghiorso. It was prepared by bombarding americium with high-speed α-particles from the Berkeley cyclotron.

$$_{95}Am^{241} + {_2}He^4 \rightarrow {_{97}}Bk^{243} + 2\,{_0}n^1 \qquad \text{(half life 4.6 hours.)}$$

This element apparently decays by orbital electron capture. A more stable isotope was also synthesized by the same team:

$$_{95}Am^{241} + {_2}He^4 \rightarrow {_{97}}Bk^{245} \qquad \text{(half life 4.95 days.)}$$

Californium (98): It was discovered in 1950 by Seaborg, Thomson, Ghiorso and Street. It was produced by bombarding microgram amounts of curium-242 with 35-Mev α-particles from the Berkeley cyclotron.

$$_{96}Cm^{242} + {_2}He^4 \rightarrow {_{98}}Cf^{244} + 2\,{_0}n^1 \qquad \text{(half life 45 min.)}$$

Less energetic α-particles produce a more stable isotope.

$$_{96}Cm^{242} + {_2}He^4 \rightarrow {_{98}}Cf^{246} \qquad \text{(half life 35.7 hours.)}$$

As the techniques were developed, heavier nuclei like carbon were tried as bullets to bombard naturally available elements like uranium to produce the very heavy elements in one step:

$$_{92}U^{238} + {_6}C^{12} \rightarrow {_{98}}Cf^{246} + 4\,{_0}n^1$$

$$_{92}U^{238} + {_6}C^{12} \rightarrow {_{98}}Cf^{244} + 6\,{_0}n^1$$

Einsteinium (99): This was prepared by bombarding uranium with nitrogen nuclei accelerated in the Berkeley cyclotron.

$$_{92}U^{238} + {_7}N^{14} \rightarrow {_{99}}En^{247} + 5\,{_0}n^1 \qquad \text{(half life 7.3 min.)}$$

Fermium (100): The discovery of element 100 was announced simultaneously by a research team at the Argonne National Laboratory and the Seaborg team at the University of California. The first production of an isotope of element 100 took place in an atomic reactor. The synthesis involved transmutation of plutonium in a complicated process. It involved absorption of 15 neutrons by each atom of plutonium and 6 β-particles are emitted in the process.

$$_{94}Pu^{239} + 2\ _{0}n^{1} \rightarrow\ _{94}Pu^{241} \rightarrow _{95}Am^{241} +\ _{-1}e^{0}$$

$$_{95}Am^{241} +\ _{0}n^{1} \rightarrow\ _{95}Am^{242} \rightarrow\ _{96}Cm^{242} +\ _{-1}e^{0}$$

$$_{96}Cm^{242} + 7\ _{0}n^{1} \rightarrow _{96}Cm^{249} \rightarrow\ _{97}Bk^{249} +\ _{-1}e^{0}$$

$$_{97}Bk^{249} +\ _{0}n^{1} \rightarrow\ _{97}Bk^{250} \rightarrow\ _{98}Cf^{250} +\ _{-1}e^{0}$$

$$_{98}Cf^{250} + 3\ _{0}n^{1} \rightarrow\ _{98}Cf^{253} \rightarrow\ _{99}En^{253} +\ _{-1}e^{0}$$

$$_{99}En^{253} +\ _{0}n^{1} \rightarrow\ _{99}En^{254} \rightarrow _{100}Fm^{254} +\ _{-1}e^{0}$$

A Swedish group of scientists at the Nobel Institute of Physics in Stockholm succeeded in making an isotope of element 100 by bombarding uranium with 180-Mev oxygen particles in a cyclotron.

$$_{92}U^{238} +\ _{8}O^{16} \rightarrow\ _{100}Fm^{252} + 2\ _{0}n^{1} \qquad \text{(half life 30 min.)}$$

Mendelevium (101): It was prepared by the Seaborg team by bombarding einsteinium coated on a gold foil with 41-Mev α-particles.

$$_{99}En^{253} +\ _{2}He^{4} \rightarrow\ _{101}Md^{256} +\ _{0}n^{1} \qquad \text{(half life 30 min.)}$$

It is unusual in that it appears to decay by spontaneous fission.

Nobelium (102): It was synthesized at the Nobel Institute of Physics in Stockholm, Sweden in 1957. It was the combined cooperative effort of research teams from three countries. Scientists from the United States Atomic Energy Commission's Argonne National Laboratory contributed the curium sample which was used as the target. English scientists from Harwell furnished carbon-13, a rare isotope, which was used as the bullet. Swedish scientists at Stockholm provided the specially constructed cyclotron required for such experiments.

$$_{96}Cm^{242} +\ _{6}C^{13} \rightarrow _{102}No^{253} + 2\ _{0}n^{1} \qquad \text{(half life 10 min.)}$$

This isotope disintegrated so rapidly that the major problem was to prove that it had been created. Only about 50 atoms of the new element were produced and special techniques were necessary to study its emissions before it disappeared completely.

Higher elements: Only a few atoms have been obtained in these cases because of the non availability of target materials and the low probability of the desired nuclear reactions.

5.6 GENERAL CHARACTERISTICS OF ACTINIDES

1) The elements are all metals with fairly high melting points though the values are considerably lower than for the transition elements.

2) The sizes of the ions decrease regularly along the series because of the poor screening of the nuclear charge by the f- orbital results in an actinide contraction similar to the lanthanide contraction.

3) The actinides are much denser than lanthanides.

4) They are reactive metals like La and the lanthanides.

5) Th shows +2, +3, +4 oxidation state. +4 is the most stable and the Th^{4+} ion is known both in the solid and in solution with basic character. $Th(NO_3)_4.5H_2O$ is the best known salt, and it is very soluble in water.

General Observations (comparisons with Lanthanides)

- Electronic Configurations of Actinides are not always easy to confirm.
- Atomic spectra of heavy elements are very difficult to interpret in terms of configuration.
- For early actinides promotion 5f $\rightarrow$ 6d occurs to provide more bonding electrons. Much easier than corresponding 4f $\rightarrow$ 5d promotion in lanthanides.
- Second half of actinide series resemble lanthanides more closely.
- 5f orbitals have greater extension with respect to 7s and 7p than do 4f relative to 6s and 6p orbitals, e.g. ESR evidence for covalent bonding contribution in UF_3, but not in NdF_3.
- 5f / 6d / 7s / 7p orbitals are of comparable energies over a range of atomic numbers, especially U–Am.
- Tendency towards variable valency. Greater tendency towards (covalent) complex formation than for lanthanides, including complexation with π-bonding ligands.
- Electronic structure of an element in a given oxidation state may vary between compounds and in solution. Often impossible to say which orbitals are being utilized in bonding.
- Ionic Radii of ions show a clear "Actinide Contraction"
- Actinide 3+ or 4+ ions with similar radii to their Lanthanide counterparts show similarities in properties that depend upon ionic radius.
- Magnetic Properties: Hard to interpret. Spin-orbit coupling is large & Russell-Saunders (L.S) Coupling scheme doesn't work. Ligand field effects are expected where 5f orbitals are involved in bonding.
- ***Survey of Actinide Oxidation States:***
-        ***+2***
  Unusual oxidation state.
  Common only for the heaviest elements.
  No^{2+} & Md^{2+} are more stable than Eu^{2+}.
  Actinide An^{2+} ions have similar properties to Lanthanide Ln^{2+} and to Ba^{2+} ions.
-        ***+3***
  The most common oxidation state.

  The most stable oxidation state for all trans-Americium elements (except No).

Of marginal stability for early actinides Th, Pa, U (But: Group oxidation state for Ac).

General properties resemble Ln^{3+} and are size-dependent.

Stability constants of complex formation are similar for same size An^{3+} & Ln^{3+}.

Isomorphism is common.

Later An^{3+} & Ln^{3+} must be separated by ion-exchange/solvent extraction.

Binary Halides, MX_3 easily prepared, & easily hydrolysed to MOX.

Binary Oxides, M_2O_3 known for Ac, Th and trans-Am elements.

- **+4**

Principal oxidation state for Th.

Th4+ chemistry shows resemblance to Zr^{4+} / Hf^{4+} - like a transition metal.

Very important, stable state for Pa, U, Pu.

Am, Cm, Bk & Cf are increasingly easily reduced - only stable in certain complexes, e.g. Bk^{4+} is more oxidizing than Ce^{4+}.

MO_2 known from Th to Cf (fluorite structure).

MF_4 are isostructural with lanthanide tetrafluorides.

MCl_4 only known for Th, Pa, U & Np.

Hydrolysis / Complexation / Disproportionation are all important in (aq).

- **+5**

Principal state for Pa.

Pa^{5+} chemistry resembles that of Nb^{5+} / Ta^{5+} - like a transition metal.

For U, Np, Pu and Am the AnO_2^+ ion is known (i.e. quite unlike Nb/Ta).

Comparatively few other An (V) species are known.

e.g. fluorides, PaF_5, NbF_5, UF_5; fluoro-anions, AnF_6^-, AnF_7^{2-}, AnF_8^{3-}.

e.g. oxochlorides, $PaOCl_3$, $UOCl_3$; uranates, $NaUO_3$.

- **+6**

AnO_2^{2+} ions are important for U, Np, Pu, Am. UO_2^{2+} is the most stable.

Few other compounds e.g. AnF6 (An = U, Np, Pu), UCl_6, UOF_4 etc..., $U(OR)_6$.

- **+7**

Only the marginally stable oxo-anions of Np and Pu, e.g. AnO_5^{3-}.

Comparison of lanthanide and actinide properties

Lanthanides	Actinides
Electronic configuration	**Electronic configuration**
The general electronic configuration of lanthanides is [Xe] $4f^{0-14}$ $5d^{0-1}$ $6s^2$	The general electronic configuration of lanthanides is [Rn] $5f^{0-14}$ $6d^{0-1}$ $7s^2$
Oxidation states	**Oxidation states**
Most common oxidation state = +3	Most common oxidation state = +3
Other oxidation states = +2, +4	Other oxidation states = +2,+4, +5, +6, +7

Atomic & ionic size	Atomic and ionic size
Size of the Ln atom or ion decreases across the period. In the group, size of a lanthanide element is smaller than the actinide of its own group.	Size of the An atom or ion decreases across the period. Actinide has largest size in its own group.
Chemical reactivity Lesser tendency towards complex formation. Except promethium, the rest are non radioactive. Lanthanides do not form oxocations They form oxides and hydroxides which are less basic.	**Chemical reactivity** Stronger tendency towards complex formation. All the actinides are radioactive. Actinides form oxocations such as UO^+, PuO^+, NpO_2^+. They form oxides and hydroxides which are more basic.
Most of the lanthanide ions are colour less	Most of the actinide ions are coloured

5.7 CHEMISTRY OF URANIUM

Uranium is a silvery-white metallic chemical element in the actinide series of the periodic table, with symbol **U** and atomic number 92. A uranium atom has 92 protons and 92 electrons, of which 6 are valence electrons. Uranium is weakly radioactive because all its isotopes are unstable (with half-lives of the 6 naturally known isotopes, U-233 - U-238, varying between 69 years and 4½ billion years).

The most common isotopes of uranium are uranium-238 (which has 146 neutrons and accounts for almost 99.3% of the uranium found in nature) and uranium-235 (which has 143 neutrons, accounting for 0.7% of the element found naturally). Uranium has the second highest atomic weight of the primordially occurring elements, lighter only than plutonium.[3] Its density is about 70% higher than that of lead, but not as dense as gold or tungsten. It occurs naturally in low concentrations of a few parts per million in soil, rock and water, and is commercially extracted from uranium-bearing minerals such as uraninite.

Uranium Occurrence: In nature, uranium is found as uranium-238 (99.2739–99.2752%), uranium-235 (0.7198–0.7202%), and a very small amount of uranium-234 (0.0050–0.0059%). Uranium is a naturally occurring element that can be found in low levels within all rock, soil, and water. Uranium is found in hundreds of minerals including uraninite (the most common uranium ore), carnotite, autunite, uranophane, torbernite, and coffinite.

Uranium Extraction: Uranium ore is crushed and rendered into a fine powder and then leached with either an acid or alkali. The leachate is subjected to one of several sequences of precipitation, solvent extraction, and ion exchange. The resulting mixture, called yellowcake, contains at least 75% uranium oxides U_3O_8. Yellowcake is then calcined to remove impurities from the milling process before refining and conversion.

Uranium Properties: Uranium generally has a valence of 6 or 4. Uranium is a heavy, lustrous, silvery-white metal, capable of taking a high polish. It exhibits three crystallographic modifications: alpha, beta, and gamma. It is a bit softer than steel; not hard enough to scratch glass. It is malleable, ductile, and slightly paramagnetic. When exposed to air, uranium metal becomes coated with a layer of oxide. Acids will dissolve the metal, but it is not affected by alkalis. Finely divided uranium metal is attached by cold water and is pyrophoric. Crystals of uranium nitrate are triboluminescent. Uranium and its (uranyl) compounds are highly toxic, both chemically and radiologically.

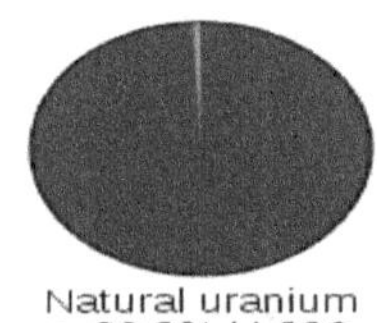

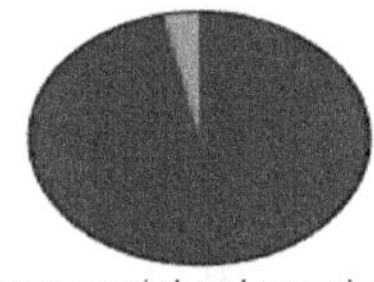

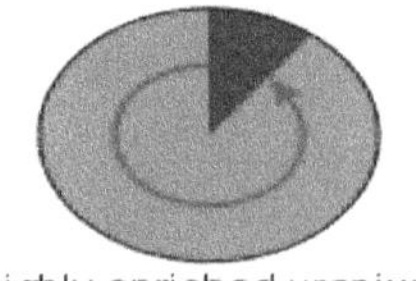

Uranium as nuclear fuel: Uranium is of great importance as a nuclear fuel. Nuclear fuels are used to generate electrical power, to make isotopes, and to make weapons. Much of the internal heat of the earth is thought to be due to the presence of uranium and thorium. Uranuim-238, with a half-life of 4.51×10^9 years, is used to estimate the age of igneous rocks. The ^{235}uranium is a fissile isotope of the element Uranium. This isotope can be split through nuclear fission, which is the process of splitting atoms or fissioning them. Other fissile isotopes are 239Pu and 232Th. The ability for an atom to fission depends upon the speed at which the neutron is moving. 232Th requires a very fast neutron to induce fission but 235U needs slower neutrons. If a neutron is too fast, it will pass right through a 235U atom without affecting it at all. The fissioning of 236U can produce over 20 different products that always add up to an atomic mass of 236.

World War II sparked the first and only time that the Atomic Bomb had been used in war to destroy the enemy. Two weapon designs were available in WWII; the gun assembly "Little Boy" A-Bomb that used 235U and the implosion assembly "Fat Man" that used 239Pu. "Little Boy" was dropped on Hiroshima, Japan on August 6, 1945 and three days later "Fat Man" was dropped on Nagasaki, Japan. These bombings ended WWII.

In a nuclear bomb, each nuclei undergoes fission releasing 2 to 3 neutrons which then all stimulate additional fission reactions that are uncontrolled. These reactions multiply exponentially and within a few milliseconds, billions of nuclei fission simultaneously, releasing a tremendous amount of energy and heat resulting in a nuclear explosion. In the gun assembly one subcritical mass (sample in such a configuration that the multiplication factor is less than 1) of 235U is driven by an explosive charge into another to create a supercritical (greater than 1) mass. In the implosion assembly using 239Pu, a bomb can only be created by surrounding a slightly subcritical sphere with explosives. When those explosives are detonated, the shock wave crushes the Pu sphere to a high enough density that it becomes supercritical and the inward motion of the Pu prevents it from flying apart before it completely reacts. Although these bombs seem easy to make, fissile 235U is almost identical to its nonfissile isotope 238U and the 239Pu bomb design is very difficult because to create an explosion that will symmetrically compress a sphere is almost impossible.

Compounds of uranium:

Oxides: The most important oxidation states of uranium are uranium (IV) and uranium (VI), and their two corresponding oxides are, respectively, uranium dioxide (UO_2) and uranium trioxide (UO_3). Other uranium oxides such as uranium monoxide (UO), diuranium pentoxide (U_2O_5), and uranium peroxide ($UO_4 \cdot 2H_2O$) also exist. The most common forms of uranium oxide are triuranium octoxide (U_3O_8) and UO_2.

❖ **UO_2 & U_4O_9**

UO_2 (black-brown) has the Fluorite structure. Stoichiometric material is best obtained from: Interstitial Oxide Ions may be incorporated into the structure - UO_{2+x}. Neutron Diffraction studies indicate oxide vacancies in the normal fluorite lattice.

$$UO_3 \xrightarrow[\text{300–600°C}]{\text{H}_2 \text{ or CO}} UO_2 + H_2O \text{ (or } CO_2)$$

At $UO_{2.25}$ (U_4O_9) (black) - interstitials are ordered forming a distinct phase in the phase diagram.

❖ **U_3O_8 & UO_3**

U_3O_8 is dark green.conveniently made by heating uranyl nitrate or ethanoate in air.

$$3UO_2(NO_3)_2 \xrightarrow[\text{650–800°C}]{\text{O}_2} U_3O_8 + 6NO_2 + 2O_2$$

> 650°C Higher uranium oxides decompose to U_3O_8.

> 800°C loses U_3O_8 oxygen.

Halides: *Fluorides*- **UF6** - the most important fluoride.

Preparation: $UO_2 + 4HF \rightarrow UF_4 + 2H_2O$.

$\qquad\qquad 3UF_4 + 2ClF_3 \rightarrow 3UF_6 + Cl_2$.

Properties: mp. 64°C, vapour pressure = 115 mmHg at 25°C. Made on a large scale to separate uranium isotopes. Gas diffusion or centrifugation separates $^{235}UF_6$ from $^{238}UF_6$. Uranium richer in 235U is termed enriched, richer in 238U is depleted. Powerful fluorinating agent.

Other Fluorides

$UF_6 + Me_3SiCl \rightarrow Me_3SiF + \frac{1}{2}Cl_2 + UF_5$ (melts to an electrically-conducting liquid).

$UF_6 + 2Me_3SiCl \rightarrow 2Me_3SiF + Cl_2 + UF_4 \rightarrow$ 500-600°C gives $UO_2 + CFCl_2CFCl_2$.

Mixed-Valence fluorides such as U_2F_9 also form.

Reduction of UF_4 with $\frac{1}{2}H_2$ yields UF_3.

❖ *Chlorides*

UCl_4 – is the usual starting material for the synthesis of other U(IV) compounds.

Preparation: Liquid-phase chlorination of UO3 by refluxing hexachloropropene.

Properties: Soluble in polar organic solvents & in water.Forms various adducts (2 - 7 molecules) with O and N donors.

UCl_3 - Usually encountered as $UCl_3(thf)x$ (a rather intractable material).

Unsolvated binary gives its name to the UCl_3 structure! Actinide trihalides form a group with strong similarities (excepting redox behaviour) to the Lanthanides.

UCl_6 - From chlorination of U_3O_8 + C. Highly oxidising. Moisture-sensitive : $UCl_6 + 2H_2O \rightarrow UO_2Cl_2$ (Uranyl Chloride) + 4HCl. In CH_2Cl_2 solution UCl_6 decomposes to U_2Cl_{10} (Mo_2Cl_{10} structure).

Organometallic compounds of uranium:

Organometallic chemistry of actinides is relatively recent. Similar to lanthanides in range of cyclopentadienides / cyclo-octatetraenides / alkyls Cyclopentadienides are π-bonded to actinides.
Compounds include:

$$UCl_3 + 3K^+(C_5H_5)^- \xrightarrow[\text{reflux}]{\text{benzene}} (C_5H_5)_3U^{III} + 3KCl$$

$$UCl_4 + 4K^+(C_5H_5)^- \xrightarrow[\text{reflux}]{\text{benzene}} (C_5H_5)_4U^{IV} + 4KCl$$

$$UCl_4 + 3Tl^+(C_5H_5)^- \xrightarrow{\text{DME}} (C_5H_5)_3U^{IV}Cl + 3TlCl$$

most important
cyclopentadienide

$C_5H_5^-$ does not behave ionically, but Cl- is labile towards formation of a wide variety of $(C_5H_5)_3UX$ compounds.

The most notable Cyclooctatetraenide is Uranocene

$$UCl_4 + 2K_2C_8H_8 \xrightarrow{\text{thf}}$$
$$U_{(g)} + 2C_8H_8 \xrightarrow{\text{MVS}}$$

- Green crystals, paramagnetic and pyrophoric- Stable to hydrolysis- Planar 'sandwich'.

- Eclipsed D_{8h} conformation.

- UV-PES studies show that bonding in uranocene has 5f & 6d contributions.

Uranium Applications:

1. The main use of uranium in the civilian sector is to fuel nuclear power plants. One kilogram of uranium-235 can theoretically produce about 20 terajoules of energy (2×10^{13} joules), assuming complete fission; as much energy as 1500 tonnes of coal.

2. Uranium may be used to harden and strengthen steel.

3. Uranium was also used in photographic chemicals (especially uranium nitrate as a toner), in lamp filaments, to improve the appearance of dentures, and in the leather and wood industries for stains and dyes.

4. Uranium salts are mordants of silk or wool.

5. Uranium metal is used for X-ray targets in the making of high-energy X-rays.

6. The long half-life of the isotope uranium-238 (4.51×10^9 years) makes it well-suited for use in estimating the age of the earliest igneous rocks and for other types of radiometric dating, including uranium-thorium dating, uranium-lead dating and uranium-uranium dating.

7. The major application of uranium in the military sector is in high-density penetrators.

8. Uranium salts have been used for producing yellow 'vaseline' glass and ceramic glazes.
9. Uranium is used in inertial guidance devices, in gyro compasses, as counterweights for aircraft control surfaces, as ballast for missile reentry vehicles, for shielding, and for x-ray targets.

5.8 CHEMISTRY OF THORIUM

Thorium is a naturally occurring radioactive chemical element with the symbol **Th** and atomic number 90. It was discovered in 1828 by the Norwegian mineralogist Morten Thrane Esmark and identified by the Swedish chemist Jöns Jakob Berzelius and named after Thor, the Norse god of thunder. In nature, virtually all thorium is found as thorium-232, which undergoes alpha decay with a half-life of about 14.05 billion years. Other isotopes of thorium are short-lived intermediates in the decay chains of higher elements, and only found in trace amounts. Thorium is estimated to be about three to four times more abundant than uranium in the Earth's crust, and is chiefly refined from monazite sands as a by-product of extracting rare earth metals. When compared to uranium, there is a growing interest in thorium-based nuclear power due to its greater safety benefits, absence of non-fertile isotopes and its higher occurrence and availability. India's three stage nuclear power programme is possibly the most well known and well funded of such efforts.

Thorium Occurrence:

Thorium is found in small amounts in most rocks and soils. Thorium occurs in several minerals including thorite ($ThSiO_4$), thorianite ($ThO_2 + UO_2$) and monazite. Thorianite is a rare mineral and may contain up to about 12% thorium oxide. Monazite contains 2.5% thorium, allanite has 0.1 to 2% thorium and zircon can have up to 0.4% thorium.

Thorium Extraction:

Thorium has been extracted chiefly from monazite through a complex multi-stage process. The monazite sand is dissolved in hot concentrated sulfuric acid (H_2SO_4). Thorium is extracted as an insoluble residue into an organic phase containing an amine. Next it is separated or stripped using an ion such as nitrate, chloride, hydroxide, or carbonate, returning the thorium to an aqueous phase. Finally, the thorium is precipitated and collected.

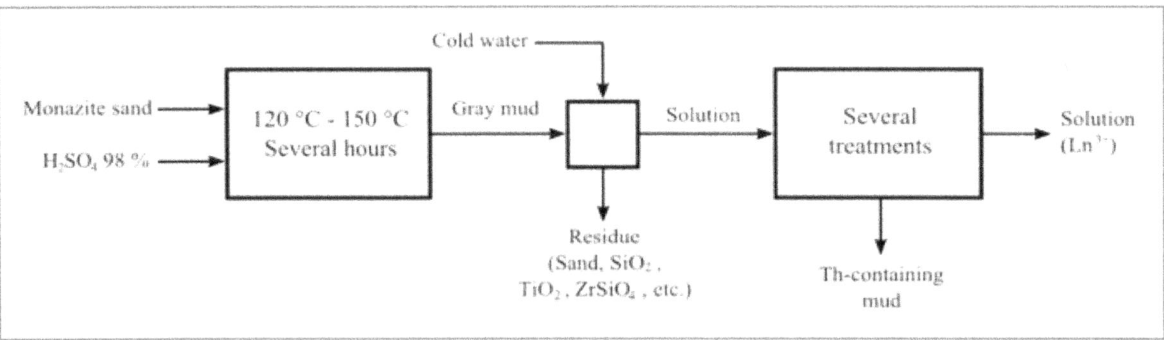

Several methods are available for producing thorium metal: it can be obtained by reducing thorium oxide with calcium, by electrolysis of anhydrous thorium chloride in a fused mixture of sodium and potassium chlorides, by calcium reduction of thorium tetrachloride mixed with anhydrous zinc chloride, and by reduction of thorium tetrachloride with an alkali metal.

Thorium properties:

Pure thorium is a silvery-white metal that is air-stable and retains its luster for several months. When contaminated with the oxide, thorium slowly tarnishes in air, becoming gray and finally black. Pure thorium is soft, very ductile, and can be cold-rolled, swaged, and drawn. Thorium has one of the largest liquid temperature ranges of any element, with 2946 °C between the melting point and boiling point. Thorium metal is paramagnetic with a ground state of $6d^27s^2$.

Thorium is slowly attacked by water, but does not dissolve readily in most common acids, except hydrochloric acid. It dissolves in concentrated nitric acid containing a small amount of catalytic fluoride ion. Thorium's oxide is ThO_2. Thorium's most common oxidation state is +4, as in ThF_4, but thorium also has an oxidation state of +3, as in ThI_3. Thorium has been shown to activate carbon-hydrogen bonds, forming unusual compounds. Thorium atoms can also bond to more atoms than any other element. For example, in the compound thorium tetrakisaminodiborane, thorium bonds to fifteen hydrogen atoms.

Thorium as nuclear fuel:

Thorium-based nuclear power is nuclear reactor-based electrical power generation fueled, ultimately, by the element thorium. According to proponents, a thorium fuel cycle offers several potential advantages over a uranium fuel cycle—including much greater abundance on Earth, superior physical and nuclear fuel properties, and reduced nuclear waste production. However, it suffers from higher production and processing costs, and lacks significant weaponization potential.

The **thorium fuel cycle** is a nuclear fuel cycle that uses the naturally abundant isotope of thorium, ^{232}Th, as the fertile material. In the reactor, ^{232}Th is transmuted into the fissile artificial uranium isotope ^{233}U which is the nuclear fuel. Unlike natural uranium, natural thorium contains only trace amounts of fissile material (such as ^{231}Th), which are insufficient to initiate a nuclear chain reaction. Additional fissile material or another neutron source are necessary to initiate the fuel cycle. In a thorium-fueled reactor, ^{232}Th absorbs neutrons eventually to produce ^{233}U. This parallels the process in uranium breeder reactors whereby fertile ^{238}U absorbs neutrons to form fissile ^{239}Pu. Depending on the design of the reactor and fuel cycle, the generated ^{233}U either fissions in situ or is chemically separated from the used nuclear fuel and formed into new nuclear fuel.

In the thorium cycle, fuel is formed when ^{232}Th captures a neutron (whether in a fast reactor or thermal reactor) to become ^{233}Th. This normally emits an electron and an anti-neutrino (v) by $\beta-$ decay to become ^{233}Pa. This then emits another electron and anti-neutrino by a second $\beta-$ decay to become ^{233}U, the fuel:

$$n + {}^{232}_{90}\text{Th} \rightarrow {}^{233}_{90}\text{Th} \xrightarrow{\beta^-} {}^{233}_{91}\text{Pa} \xrightarrow{\beta^-} {}^{233}_{92}\text{U}$$

A **breeder reactor** is a nuclear reactor capable of generating more fissile material than it consumes. These devices are able to achieve this feat because their neutron economy is high enough to breed more fissile fuel than they use from fertile material like uranium-238 or thorium-232. India has the capability to use thorium cycle based processes to extract nuclear fuel. This is of special significance to the Indian nuclear power generation strategy as India has one of the world's largest reserves of thorium, which could provide power for more than 10,000 years , and perhaps as long as 60,000 years.

Thorium applications:

1. Thorium is a component of the magnesium alloy series, called Mag-Thor, used in aircraft engines and rockets and imparting high strength and creep resistance at elevated temperatures.
2. Thoriated magnesium was used to build the CIM-10 Bomarc missile.
3. Thorium is also used in its oxide form (thoria) in gas tungsten arc welding (GTAW) to increase the high-temperature strength of tungsten electrodes and improve arc stability.
4. In electronic equipment, thorium coating of tungsten wire improves the electron emission of heated cathodes.
5. Uranium-thorium age dating has been used to date hominid fossils, seabeds, and mountain ranges.
6. Thorium dioxide is a material for heat-resistant ceramics, e.g., for high-temperature laboratory crucibles.

7. Thorium dioxide has been used as a catalyst in the conversion of ammonia to nitric acid, in petroleum cracking and in producing sulfuric acid.
8. Thorium fluoride (ThF_4) is used as an antireflection material in multilayered optical coatings. It has excellent optical transparency in the range 0.35–12 μm.

5.9 EXERCISE QUESTIONS

Exercise Questions: Type 1
1. What are inner transition elements?
2. Give the general electronic configuration of actinide elements.
3. Decide which of the following atomic numbers the atomic numbers of inner transition elements are: 39, 59, 74, 95, 102, 104.
4. Name the members of the lanthanide series which exhibit +4 oxidation state?
5. Mention an important ore of uranium.
6. What is meant by a breeder reactor?

Exercise Questions: Type 2

1. Explain the extraction of lanthanide metals from the monazite ore.
2. Explain briefly the lanthanide contraction.
3. Explain the synthesis of actinide metal: Neptunium.
4. Actinide contraction is greater from element to element than lanthanide contraction. Justify.
5. Write a short note on thorium fuel cycle.

Exercise Questions: Type 3

1. Compare the chemistry of actinoids with that of the lanthanides with special reference to:
   i. Oxidation state
   ii. Atomic and ionic size
   iii. Chemical reactivity
2. Explain the applications of the following actinide metals in various sectors,
   i. Uranium
   ii. Thorium
3. Which is the last element in the series of the actinides? Write its electronic configuration. Comment on the possible oxidation states of this element.

CURRICULUM VITAE OF THE AUTHORS

1. 0 Dr. H.C. Ananda Murthy

1.1 Personal Information

Name   : Dr. H.C. Ananda Murthy

Date of Birth   : 06-07-1973 Age: 44 years   Gender: Male

E-mail Id & Contact No.   anandkps350@gmail.com, anandkps7@yahoo.com

                     & +255 768 406 138

Nationality   : Indian

Marital Status   : Married Number of Children: 2

1.2 A: Academic Information (Completed studies at Government Universities in India)

Qualification	Descipline/ subject	School/college/ University	Year of passing	Class with Percentage
B.Sc.	PCM	Bangalore University	1995	First Class 60.16%
M.Sc.	Chemistry (Inorganic Chemistry)	Bangalore University	1997	First Class 66.60%
M.Phil.	Chemistry (Composite materials)	Gharathidasan University, Tirchy	2996	First Class 76.40%
Ph.D.	Chemistry (Electrochemistry of composite materials) Physical Chemistry)	VT University, Kamataka	2012	

1. 2.B: Title of the Ph.D. Degree

A Study on the effect of Ceramic reinforcement on the corrosion behavior of Aluminium metal matrix composites.

1. 2.C: Major Courses Studied

SL.No.	Degree	Couses Leaned
1.	B.Sc	Physics, Chemistry and Mathematics
2.	M.Sc	Coordination Chemistry, bio-Inorganic Chemistry, Organometallic Chemistry & Catalysis, Molecular Spectroscopy, Solid State Chemistry, Industrial and Nuclear Chemistry, Analytical Chemistry.
3	M.Phil.	Apectroscopy, Inorganic Chemistry, Research Methodology, Corrosion Theory.
4	Ph.D	Corrosion Science, Mateial Science, Instrumental Methods of Analysis, Research Methodology.

1.3 A: Teaching Experience: Total 19 years

Period	Institution	Designation	Total Year of teaching
Before Ph.D (1997-2012)	Vivesvaraya Technological University, India	Lecturer (9 years) Senior Lecturer (3 years) Asst. Prf. & HOD (3 years)	15 years
After Ph.D (2012-2016)	Vivesvaraya Technological University, India The University of Dodoma, Tanzania	Asst. Prof. & HOD (1 year) Lecturer (3 Years)	4 years

1. 3. B: Major Courses Taught for U G Progam (Taught 19 years)

1. Coordination and Organometallic Chemistry
2. Transition Elements Chemistry
3. Organomitallic Chemistry
4. Electrochemistry and Corrosion Protection
5. General Chemistry
6. Engineering Chemistry

1. 3.C: Major Courses Taught for P G Progam (Taught for 1 year)

1. Advanced Coordination Chemistry
2. Surface Chemistry and Corrosion Science
3. Advanced Spectroscopy
4. Advanced Environmental Chemistry
5. Theoretical and Applied Electrochemistry

1.4 A: Scientific Research Publications: Total Number of Publications = 17
National and International Journal Papers = 13, Conference Papers = 4

SL.No.	Title of the Published Papers	Cooperated authors	Jornal (Volume, Issues, date)
1	Adsorption of Mercury (II) from Aqueous Solution Using Gum Acacia-Silica Nanocomposite: Kinetics, Equilibrium and Thermodynamics Studies	Somit Kumar Singh Vandana Singh	Advanced Material letters VBRI Press, Sweden -2016 (Accepted for publication in March, 2016)
2	Electoanalytical study on the corrosion behavior of TiO_2 particulate reinforced Al 6061 composites	Single author	Material Science Research India, Vo. 12(2), 112-126 (2015). Doi.org/10.13005/msri/1205
3	Influence of TiC particulate reinforcement on the corrosion behavior of Al 6061 Metal matrix composites	Somit K. Singh	Advanced Materials letters VBRI Press, Sweden – 2015, 6 (7), 633-640.doi: 10.5185/amlett.2015.5654

4	Effect of TiN particulate reinforcement on corrosion behavior of Aluminium 6061 composited in chloride medium	V Bheemaraju C Shiva Kumara	Bulletin of Material Science-Vol.36,No.6,Nov. 2013, pp. 1057. Indian Academy of Sciences (Springer)
5	Electrochemical investigation of corrosion inhibition of AA6063alloy in IM hydrochloric acid using Schiff base compounds	V Bheemaraju P F Sanaulla H B Lokesh	IOSR Journal of Applied Chemistry (IOSR-JAC) ISSN: 2278-5736. Vlume 2, Issue 5 (Nov.-Dec. 2012), Pp 37-47
6	Electrochemical Behaviour of AA6063 Alloy in Hydrochloric Acid using Schiff Base Compounds as Corrosion Inhibitors	V Bheemaraju P F Sanaulla H B Lokesh	International Journal of Engineering Research and Applications (IJERA) ISSN: 2248, Vol. 2, Issue 5, 2012.
7	Corrosion inhibition of AA 6061 and AA6063 alloys in hydrochloric acid media by Schiff base compounds	V Bheemaraju P F Sanaulla	Journal of Chilean Chemical Society, 57, N° 4 (2012)
8	Corrosion Characterization of TiO_2 Particulate Reinforced Al-6063 Composites in Chloride and Nitrate Media	V Bheemaraju P F Sanaulla	Asian Journal of Chemistry, ISSN 0970-7077, 2011, Volume 23, Issue 4, 1664 – 1668.
9	Influence of chlorides, nitrate and sulphate media on corrosion behaviour of TiO_2 particulate reinforced Al-6061 composites	V Bheemaraju P F Sanaulla	PortugaliaeElectrochimicaActa, ISSN 1647-1571, *2010, 28(5), 309-320*
10	Corrosion behaviour of TiO_2 particulate reinforced Al-6063 composites in sodium sulphate medium	V Bheemaraju P F Sanaulla	International Journal of Applied Chemistry, ISSN 0973-1792, Vol. 6, 2, (2010) p. 225
11	A Study on the effect of TiO_2 particulate on the corrosion behaviour of Al-6063 based composites	V Bheemaraju H B Lokesh	Material Science, An Indian Journal, MSAIJ, 4(4), June 2008 [328-334]
12	Estimation of copper in brass by iodo-potentiometric technique of analysis	O G Palanna	*Analytical Chemistry*, An Indian Journal, ACAIJ, 7(9), August 2008 [694-702]

| 13 | Estimation of copper in copper salts at different pH conditions and buffer media by iodo-potentiometric technique of analysis | O G Palanna | *Analytical Chemistry*, An Indian Journal, ACAIJ, 7(9), July 2008 [684-693] |

1.4. B: Paper Presentations in National/International Conferences

14 **Anand Murth H. C, Kusuma, ShobhaBhaskara, Bheema Raju V**

Corrosion behavior of 6061 Al and its TiB_2 particulate reinforced composites in chloride medium, International Conference on Electrochemical Science and Technology (ICONEST -2014), August 7-9, ECSI, Bengaluru, India.

15. **H.C. Ananda Murthy,** Suba Rameshand V. BheemaRaju Evaluation of corrosion behaviour of TiO_2 particulate reinforced Al-6061 composites by polarisation techniques, **National Conference on Recent Advances in Chemical and Environmental Sciences (NCRACES- 2011)** held on 28 & 29th of Dec 2011.

16. **H.C. Ananda Murthy**, O.G.Palanna, T. Jeevananda, Subaramesh, RamanathPrabhu, N. Suma A Novel iodo-Potentiometric method of estimation of percentage of MnO_2 in Pyrolusite, **National Conference on Recent Trends in Materials Chemistry and Engineering (RTMCE-2011)**, September 29-30, 2011, Department of Chemistry, RNS Institute of Technology, Bangalore 560098.

17. **H.C. Ananda Murthy** and V. BheemaRaju,

Corrosion behaviour of TiO_2 particulate reinforced Al-6061 in chloride, nitrate, sulphate and acidic media, International conference on Recent Advances in Industrial Electrochemical Science and Technology (ICRAIEST-2009), November 2009, Department of Chemistry, Mangalore University, Mangalagangothri, Mangalore.

2. Seminars and Training Programmes

Seminars/Course	Duration	Organisers
Workshop on "Research Methodologies and Report Writing	2 days	Visvesvaraya Technological University, Mysore on May 28th and 29th 2009
TEQUIP sponsored Short Term Training Programme	5 days	BMS college of Engineering And Dr. A I T, Bangalore from 26th 2006.
Workshop on "Industrial Corrosion Awareness & Prevention"	1 day	Mangalore University, Mangalagangothri, Mangalore on 25th February 2006.
AICTE-ISTE Sponsored short tern training programme	5 days	from 23rd to 27th August 2004 at Dr. Ambedkar Institute of Technology, Bangalore.

3. Written Books,Book Chaper and Compendia

Title (Written)	Publisher	Year
1. Corrosion Science for University Students	Lulu Press, Inc. 3101 Hillsborough St, Raleigh, NC 27607. ISBN: 978-1-329-97036-6	2016
2. Corrosion Behaviour of Reinforced Aluminium MMCs	LAMBERT Academic Publishing, Germany. ISBN: 978-3-659-56260-0	2014
3. Electrochemistry Chemistry and Corrosion protection (CH 314) for 3rd Year B.Sc. course.	The university of Dodoma, Tanzania.	2014
4. General Chemistry (CH 111) for 1st Year B.Sc. course.	The university of Dodoma, Tanzania.	2014
5. Coordination chemistry and Organo-metallic chemistry (CH 221) for 2nd Year B.Sc. course.	The university of Dodoma, Tanzania.	2014
6. Organometallic chemistry (CH 310) for 3rd year B.Sc. course.	The university of Dodoma,Tanzania.	2014
7. Engineering Chemistry	Wiley-India ISBN: 978-81-265-1988-0	2011

1.5 Details of Scholarship/Awards/Honours

1. Prathibha puraskar-2012 by Devanga Employees Association, Bangalore.

2. NagarikaSanman- 2012 by Dr AIT employees welfare association, Bangalore.

1.6 Passport Details

Passport Number: K8778065
Date of Issue: 17/09/2012 Place of Issue: BANGALORE
Date of Expiry: 16/09/2022   Passport Validity: 10 years

1.7 DETAILS OF EXTRA-CURRICULAR ACTIVITIES

The Innovation: JAL-BHOOMI -THE AMPHIBIOUS CYCLE

I have successfully completed a project titled "Amphibious bicycle" sanctioned by National Innovation Foundation of India (NIF) with Three students of 7th semester Electronics and communications Engineering Branch in the year 2008 in RNS Institute of Technology. This Amphibious bicycle was displayed in VahanYatra held at Karnataka Trade Promotion Centre, Near ITPL, Bangalore in December 2008.

The amphibious bicycle comprises of a conventional bicycle modified with two extra attachments that enables it to run on both water and land.

I certify that the information given above is true, complete and correct to the best of my knowledge and belief.

Place: Dodoma, Tanzania.

2. Prof. Said Ali Hamad Vuai

1. BIODATA

1.1 Name	: Said Ali Hamad Vuai
1.2 Date ofbirth	: 09/11/1972
1.3 Sex	: Male
1.4 Place of Birth	: Kitambuu, Pemba
1.5 Nationality	: Tanzanian
1.6 Marital status	: Married

2.0 EDUCATIONBACKGROUND

(Start with the most recent awards)

2.1. June2013-April2014. University of Siena, Siena, Italy. Post-Doctoral on Environmental Sustainability Assessment

2.2.Oct. 2001-Sept.2004. University of the Ryukyus, Nishihara-cho, Okinawa, Japan. PhD in Marine and Environmental Sciences (Chemistry major).

2.3. Oct.1999-Sept. 2001,University of the Ryukyus, Nishihara-cho, Okinawa, Japan M Sc. in Chemistry, Biology and Marine Sciences (Chemistry major)

2.4. Oct.1993-June1997.University of Dar es Salaam, Tanzania B Sc. in Chemical and Processing Engineering.

2.5. Jan.1989-Nov.1992 Fidel Castro Secondary School, Vitongoji, Pemba, Zanzibar Certificate of Secondary Education & Advance certificate of Secondary Education

2.6. Jan. 1978- Nov. 1988 Mchangamdogo School, Mchangamdogo, Pemba Zanzibar

3.0 WORKEXPERIENCE/EMPLOYMENT

3.1 Record at the University of Dodoma

3.1.1. 10th April2013 to date Associate Professor, College of Natural and Mathematical Sciences

3.1.2. 6th Oct. 2008-9thApril, 2013. Senior Lecture, College of Natural and Mathematical Sciences.

3.1.2. 4th Dec. 2008-30th June,2011. Head of Department of Physical Sciences

3.1.3 1stJuly. 2011 to date Dean, School of Mathematical Sciences

3.2 Work experience elsewhere

3.2.1   Dec. 2004– Oct. 2008   Lecturer Department of Sciences, The State University of Zanzibar (SUZA).

3.2.2.March. 1999-Sept. 1999 Quality Control Manager   Fahari Beverages Ltd   Bokoba Branch.

4.0   MEMBERSHIP IN COMMITTEES/LOCAL AND INTERNATIONAL BODIES/ORGANIZATIONS.

4.1   University Committee/Boards

4.1.1 University of Dodoma Senate   since July2011

4.1.2 University of Dodoma Advisory Committee   since July 2011

4.1.3 College of Natural and Mathematical Sciences since July 2011

4.1.4 Board of School of Mathematics Sciences   since July 2011

4.1.4 Board of School of Physical Sciences   since July 2011

4.1.6 University of Dodoma Human Resource Committee   since July2011

4.1.5 Board of School of Natural Sciences and Mathematics   2009-2011

4.1.6 Senate of Research and Publication   2009-2010

4.2   National Committee/Boards:

4.2.1 National Stakeholder advisory Forum for Africa Stockpile Program

4.2.2 Public Health Research Group

4.3   International Committees/Board:

4.3.1. West Indian Ocean Marine Research Association (WIOMSA since2004

4.3.2. Society of Environmental Toxicology and Chemistry(SETAC) since2006

5.0   RESEARCH PROFILE

Date	Theme	Funding Agency
5.1. 2006-2008	Impact of Tourism on Child and Zanzibar	CODESRIA
5.2. 2006-2011	Socio-economic and ecological impact of Vicres periodic sedimentation of the livelihood of Flood plain Communities in the Lake Victoria Basin	COSTECH
5.3.2011	The potential of native Gracilaria seaweed an alternative cheap source of agar for tissue culture of vetetatively propagated crops in Tanzania	COSTECH

6.0   STUDENT'S SUPERVISION

6.1   Ali H.R. (2009). Determination of Chemical Composition of Cave water: A case study of Zanzibar Island, Tanzania. M. SC. Dissertation. Makerere University, Uganda.

6.2   Hamad B.A.(2010) Nutrients dynamic in Jozani Forest Reserve, Zanzibar.

M Sc. Dissertation. University of Dodoma6.3   Juma Mohammed (2011) Socio-economic and ecological importance of Bwawani Pond, Zanzibar. M Sc. Dissertation, University of Dodoma

6.4 Getachew R. Assessing domestic in water harvesting: Reliability, quality and point-of-use treatment systems in semi-arid city of Mekelle, Ethiopia. PhD Dissertation.

6.5 MorrisF.Hiji.AstudyonprocessofproductionofSulfuricacidfromSamena pyrites.

7.0 PUBLICATIONS
7.1. Publications in recognized/referred Journals

7.1.1 Abdallah H. Mtumwa , Edwin Paul and Said A.H. Vuai (2015) Determinants of under nutrition among Women in the Reproductive Age Group in Mainland-Tanzania. *South African Journal of Clinical Nutrition* (Accepted).

7.1.2 Fabrizio Saladini, Said A. Vuai, Benard Langat, Mathias Gustavsson, Richard Bayitse, Andrew Gidamis, Belmakki Mohammed, Amal Owis, Konanani Rashamuse, Daniel Sila, Simone Bastianoni (2015) Sustainability assessment of selected biowastes as feedstocks for biofuel and biomaterial production by emergy evaluation in five African countries. *Biomass & Bioenergy* (Accepted).

7.1.3 Getachew Redae Taffere, Yilma Seleshi, Abebe Beyene, Said A.H. Vuai, Janvier Gasana and Yilma Seleshi (2015)Location and time-specific investigation of roof rainwater quality is important to safeguard public health. Desalination and Water Treatment DOI: 10.1080/19443994.2015.1107757.

7.1.4Getachew Redae Taffere, Yilma Seleshi, Abebe Beyene, , Said A.H. Vuai, Janvier Gasana and Yilma Seleshi (2015) Reliability analysis of roof rainwater harvesting systems in a semi-arid region of sub-Saharan Africa: case study of Mekelle, Ethiopia. *Hydrological Sciences Journal* DOI:10. 1080/02626667.2015.1061195.

7.1.5 Said A. H. Vuai (2015) Aluminum, Silicon and Nutrients Characteristics in Precipitation of Semi-Arid Area in Dodoma Municipality, Tanzania. *International Journal of Environment and Toxicological Research* 92-100.

7.1.6 Morris Frank Hiji, Justin William Ntalikwal, Said Ali Vuai (2014)Producing sulphuric acid in Tanzania and potential sources: A review. *American Journal of Chemistry and Applications*1(4): 40-44.

7.1.7 Said S. Bakari, Per Aagaard, Rolf D.Vogt, Fridtjov Ruden, Ingar Johansen and **Said Ali Vuai.** (*2013)* Strontium isotopes as tracers for quantifying mixing of groundwater in the alluvial plain of a coastal watershed, South-eastern Tanzania. *Applied Geochemistry,*(accepted).

7.1.8**Vuai, S. A. H.**, Mungai, N. W.and Ibembe, J. D.(2012) Effect of Land Use Activities on Spatial and Temporal Variation of   Nutrients deposition in Mwanza Region: Implication to the Atmospheric loading to the Lake Victoria. *Atmospheric and Climate Sciences*(in press).

7.1.9 Hasan Rashid Ali, Nyakairu J. and **Said A. Vuai** *(2012)*. Influence of tides to the cave water quality: A case study   on Zanzibar Island, Tanzania. *International Journal of Current Research*Vol.4, Issue, 04,pp.327-334.

7.1.10 **Vuai, S. A. H.,** Mungai, N. W. and Ibembe, J. D.*(2012)*Spatial variation of Nutrients in Sondu-miriu and Simiyu-Duma Rivers: Implication on sources and factors influencing their transportation into the Lake Victoria. *Journal of Earth Sciences and Climate Change. 3:119.* doi:10.4172/2157-7617.1000119

7.1.11 **Vuai, S. A. H** *(2012)*. Microbialand nutrient contamination of domestic well in Urban-West Region, Zanzibar, Tanzania. *Airand Water Born Diseases,*1:102. doi:10.4172/2167-7719.1000102.

7.1.12 Said Suleiman Bakari, Per Aagaard, Rolf D. Vogt, Fridtjov Ruden, Ingar Johansen and **Said Ali Vuai** (2012). Delineation of groundwater proven an ceina coastal aquifer using statistical and isotopic methods, Southeast Tanzania. *Environmental Earth Sciences: 1-14.*

7.1.13 Ali, H.R., Nyakairu, G.W. and **Vuai, S. A. (2012**) Determination of Chemical Composition of Cave water: A case study of Zanzibar Island, Tanzania International Journal of Current Research. 4:327-334, 2012.

7.1.14 Nancy W. Mungai, Njue A. M., Abaya Samuel G.,**Vuai Said A. H**. and Ibembe John D. *(2011)* Periodic flooding and land use effects on soil properties in Lake Victoria basin *African Journal of Agricultural Research Vol. 6(16),*4613-4623.

7.1.15 JohnDanielBakibinga-Ibembe1,**VuaiA.Said**andNancyW.Mungai.(2011) Environmental laws and policies related to periodic   flooding and sedimentation in the Lake Victoria Basin (LVB) of East Africa. *African Journal of Environmental Science and Technology Vol.5(5),*367-380.

7.1.16 **Said Ali Hamad Vuai** and Tokuyama,A.   (2011) Trend of trace metals in precipitation around Okinawa Island, Japan. *Atmospheric Research*, 99, 80-84.

7.1.17 **Said Ali Hamad Vuai**. (2010). Characterization of MS Wand related waste- derived composition Zanzibar municipality. *Waste Management and Research*, 28, 177-184.

7.1.18 **Said Ali Vuai** and Akira(2007). Tokuyama. Solute generation and CO_2 consumption in the silicate rock area under subtropical climatic condition. *Chemical Geology* 236, 199–216.

7.1.19 Mohamed MkadamKombo, Tetsuya Yonaha, **Said Ali Vuai** and Akira Tokuyama (2005). Dissolved aluminum and silica release on the interaction of Okinawan subtropical red soil and seawater at different salinities: Experimental and fieldobservations. *Geochemical Journal,Vol.* 61pp. 591-601.

7.1.20 Mohamed Mkadam Kombo, **Said Ali Vuai**, Maki Ishiki and Akira Tokuyama (2005). Influence of salinity on pH and aluminum upon the interaction of acidic red soil with seawater.*JournalofOceanography,* Vol. 61, pp. 591-601.

7.1.21 **Said Ali Vuai**, Kazuyo Nakamura and Akira Tokuyama (2003). Geochemical characteristics of run of from acid sulfate soils in northern area of Okinawa Island, Japan, *Geochemical Journal*, Vol. 37, pp. 579-592.

7.1.22 **Said Ali Vuai,** Maki Ishiki and Akira Tokuyama. Acidification of fresh water by red soils in subtropical silicate rock area, Okinawa, Japan (2003). *Limnology*, Vol. 4. pp. 63-71.

7.1.23 Said Suleiman Bakari, **Said Ali Vuai** and Akira Tokuyama (2004). Forest soil fertility in Okinawa Island, Subtropical Region in Japan. *Bulletin of Faculty of Science, University of the Ryukyus,* Vol. 78, pp.276-290.

7.1.24 Said Suleiman Bakari, **Said Ali Vuai**, F.L.Manuzi and Akira Tokuyama (2004). Assessment of Self-purification of Kizinga River-Dar es Salaam, Tanzania. *Bulletin of Faculty of Science, University of the Ryukyus,*Vol.78, pp.291-310.

7.1.25 Mohamed Mkadam Kombo, **Said Ali Vuai** and Akira Tokuyama(2003). Effect of ionic strength on aluminum and dissolve silicaunderred soil- seawater interaction. *Geochimicaet Cosmochimica Acta*, Vol.67, No18 (S1), pp. A230.

7.1.26 **Said Ali Vuai**, Kazuyo, Arakaki and Akira Tokuyama (2003). Dissolved Al, Fe, Mn, Cuand Znin surface and ground waters from the northern area of Okinawa Island, *Geochimicaet Cosmochimica Acta*, Vol.67, No18 (S1), pp. A 516.

7.1.27 Mohamed Mkadam Kombo, **Said AliVuai** and Akira Tokuyama(2003). Impact of acidic red soil on pH and aluminum under the soil-seawater interacting environment. *Bulletin of Faculty of Science, University of the Ryukyus*, No. 75 pp. 75-87.

7.1.28 **Said Ali Vuai**, Akira Tokuyama, Y. Tokashiki, and Moritaka Shimo, (2001). Particle size distribution, mineral sand chemical composition of red soil from silicate rock area of Central Okinawa. *Bulletin of Faculty of Science, University of the Ryukyus*, No. 71 pp. 71-77.7.1.25 **Said Ali Vuai**, Mariko Yonashiro and Akira Tokuyama (2000). Effect of Ionic strength on the absorbance in aluminum-Tiron complex. *Bulletin of Faculty of Science, University of the Ryukyus,* No. 70 pp. 97-102.

7.2 Publications in edited proceedings:

7.2.1 Bakibinga-Iembe, J.D., Mungai, N.W and Said A. Vuai (2011). Effect of Periodic flooding and sedimentation on Community Liveligoods in Sondu- Miriu Flood plain, Kenya, *Natural Resources Management and Land Use, Proceedings of the Cluster Workshop, Inter-University Council for East Africa, LakeVictoria Research Initiative. ISBN978-9970-452-02-6*

7.2.2 Ouma, K.O., Mungai, N.W., Said A. H. Vuai and Bembe, J.D.(2011) Spatial-Temporal Vairation of Nitrogen, Phosphorous and Sediments in surface run off and adjacent river system in Sondu-Miriu Bains, Kenya. *Natural Resources Management and Land Use, Proceedings of the Cluster Workshop, Inter- University Council for East Africa, Lake Victoria Research Initiative. ISBN978-9970-452-02-6*

7.3   Publication in Books

7.4   Thesis and dissertations:

7.4.1 Influence of acidic red soil and acid sulfate soil on the quality of fresh water located in the silicate rock area, northern Okinawa Island.

PhD Thesis, University of the Ryukyus, Okinawa, Japan, September, 2004.

7.4.2 Effect of red soil on chemical composition of natural waters.
Master's thesis, University of the Ryukyus ,Okinawa, Japan, September, 2001.

7.4.3 Design, construction and evaluation of non-convecting, saltgradient solar pond for storage of thermal energy.
Graduationthesis, University of Dar es Salaam ,Tanzania, June 1997

7.5   Student compendia

7.5.1 Coordination Chemistry: teaching Compendium for Structural Inorganic Chemistry (CH 202) and Inorganic Chemistry II (CH 330) 2010.

7.5.2   Basic principles of inorganic Chemistry: Teaching compendium for Basic inorganic Chemistry

Presentations in scientific conferences and workshops:

7.6.1 **Vuai, S. A. H.,** Mungai, N. W. and Ibembe, J. D. *Socio-economic and Ecological impacts of periodic sedimentation on the livelihoods of flood plain communities of Lake Victoria.* Lake Victoria Research (Vic Res) Initiative *Biennial* Forum from 9th to 13th October, 2011, Monyonyo, Kampala, Uganda.

7.6.2 **Vuai, S.A. H**. *Spatial and Temporal variation of nutrient along Simiyu-Duma and Sondu Miriu River*. Land Use and Natural Resource Cluster workshop, 30th October to 2nd November, 2011, Nairobi, Kenya.

7.6.3 **Vuai, S.A.H**. Spatial Variation of nutrient pollution in the Lake Victoria Basin. *Stakeholders Workshop for Vicres Projec*t December 7th 2011, Hotel Barbados, Kampala. 6.4 **Vuai S. A**., Ali, H.R. and Bakari, S.S. *Geochemical Characteristics of Sapaleotherms and Water in Cave from Zanzibar Island, Tanzania*. Int. Conf. on Geosciences for Global Development (Geo Dev), Dhaka, Bangladesh, October 26-31, 2009. Abstract Volume, 84 pp.

7.6.5 Bakari, S.S., Aagaard, P., Vogt, R.D., Ruden, F. and **Vuai, S.A**. *Geochemical Characterization of the Upper and Lower Aquifer System in the Coastal Area, Tanzania*. Int. Conf. on Geosciences for Global Development (Geo Dev), Dhaka, Bangladesh, October 26-31, 2009. Abstract Volume, 63 pp.

7.6.6 **Said A. H. Vuai**, Mohammed A. Sheikh and T. Oomori. Occurrence of tributylt in (TBT)compounds in the major ports of Tanzania. International conference on pesticide use in developing countries: environmental fate, effects and public health implications, Arusha,16-20 October 2006.

7.6.7 Kombo M., Tetsuya Y., **Vuai, S**. and Tokuyama A. Effect of ionic strength on dissolved silica in the interaction of acidic red soils from Okinawa Island and seawater solutions. 51st Annual Meeting of Geochemical Society of Japan, Shizuoka University, September 20-22, 2004.

7.6.8 SaidSuleimanBakari,**SaidAliVuai**andAkira Tokuyama. Influence of underlain rock on the dissolved organic nutrients and metal release in the forest soil in Okinawa Island, Japan. 51st annual meeting of Geochemical Society of Japan, Shizuoko University, September 20-22, 2004.

7.6.9 **Said Ali Vuai**, Kazuyo Arakaki and Akira Tokuyama. Dissolved Al, Fe, Mn, Cu and Zn in surface and ground waters from the northern area of Okinawa Island. 13th V. M. Gold schimidt International Conference, Kurashiki, Japan September 7-12, 2003.

7.6.10 MohamedMkadamKombo,**SaidAliVuai**,MakiIshikiandAkira Tokuyama.

Influence of salinity on pH and aluminum up on the interaction of acidic red soil with seawater solutions. 4th International workshop on the oceanography and fishery in the East China Sea. University of the Ryukyus, Okinawa, Japan. November, 8-9, 2003.

7.6.11 Mohamed Mkadam Kombo, **Said Ali Vuai** and AkiraTokuyama. Effect of ionic strength on aluminum and dissolved silica underred soil-seawater interaction.13th V.M. Goldschimidt International Conference, Kurashiki, Japan, September 7-12, 2003.

7.6.12 **Said Ali Vuai** and Akira Tokuyama. Characteristic sofrun off from acid sulfate soils in the northern area of Okinawa Island. 49th   Annual Meeting of Geochemical Society of Japan, Kagoshima University, September 25-27, 2002.

7.6.13 Mohamed Mkadam Kombo, **Said Ali Vuai** and Akira Tokuyama. Effect of acidic red soil on pH and chemical composition of seawater. 49th Annual Meeting of Geochemical Society of Japan, Kagoshima University, September 25-27, 2002.

7.6.14 AkiraTokuyama, Mariko Yonashiro and **Said Ali Vuai**. 47   Annual Meeting of Geochemical Society of Japan,Yamagata University, Yamagata, Japan, September 25-27, 2002.

7.7   Other writings including consultancy reports and extension materials/Submitted Manuscripts:

7.7.1. Getachew Redaie Taffere, Abebe Beyene Hailu, Said Vuai and Janvier Gassana and Yilma Sileshi. Reliability analysis of roof rain water harvesting systems in semi-arid area of Mekelle City, northern Ethiopia. Journal of Hydrology (2015).

7.7.2. Fabrizio Saladini, Said A. Vuai, Benard Langat, Mathias Gustavsson, Richard Bayitse, Andrew Gidamis, Belmakki Mohammed, Amal Owis, Konanani Rashamuse, Daniel Sila, Simone Bastianoni. Sustainability assessment by means of emergy evaluation of selected biowaste as feed stocks for biofuels and biomaterials production in five African countries. Biomas Bioenergy (2015).

REFERENCES

1. Housecraft, C.E. and Sharpe, A.G. Inorganic Chemistry, Prentice Hall, Second Edition 2004.
2. Douglas, B.E., Mcdaniels, D.H. and Alexander, J.J. Concepts and Models in Inorganic Chemistry, Wiley 1994.
3. Purcell, K.F. and Kotz, J.C., Inorganic Chemistry, Prentice Hall 1980.
4. Elschenbroich, C. and Salzer, A. Organometallics, A Concise Introduction, Wiley VCH Verlag, 2nd edition 1992.
5. House, J. Inorganic Chemistry, Elsevier Science and Technology 2008.
6. Elschenbroich, C. and Salzer, A. Organometallics, A Concise Introduction, Wiley VCH Verlag, 2nd edition 1992.
7. M. Gerloch, E.G. Constable, Transition Metal Chemistry, VCH Verlagsgesellschaft mbH, D-69451 Weinheim (Federal Republic of Germany), 1994.
8. S.F.A.Kettle, Physical Inorganic Chemistry, Spectrum, 1996.
9. R Gopalan and V. Ramalingam, Concise coordination chemistry, Vikas publishing house Pvt Ltd, 2014.
10. Electronic (Absorption) Spectra of 3d Transition Metal Complexes
    S. Lakshmi Reddy, Tamio Endo and G. Siva Reddy
11. ChandraShekara Reddy, Advanced Coordination Chemistry, Campus Books International, India, 2015.
12. P.S.Sindhu, Elements of Molecular Spectroscopy, New Age International Publishers, India, 2010.
13. R.C. Mehrotra, Organometallic chemistry, New Age International publishers, 2015.
14. Robert H Crabtree, The organometallic chemistry of transition elements, Wiley-Interscience, 2005.
15. Perez, N. Electrochemistry and Corrosion Science. Kluwer Academic Publishers, 2004.
16. Mars G Fontana, Corrosion Engineering Tata McGraw Hill, 2005
17. Engineering Chemistry, John Wiley International, 2010.
18. Koryta, J. Principles of Electrochemistry, 2nd ed. J. Wiley & Sons, 1993.
19. Plieth, W. Electrochemistry for Material Science. Elsevier B.V. 2008.
20. Comninellis, C. and Chen, G. Electrochemistry for the Environment. Springer, 2010.
21. H C Ananda Murthy, Corrosion Behaviour of Reinforced Al MMCs, Lambert Academic Publishing, Germany, 2014.
22. H C Ananda Murthy, Corrosion Science, Lulu Press Inc, USA, 2016.
23. Vitousek, P. M., Mooney, H. A., Lubchenco, J., and Melilli, J. M. Human domination of Earth's ecosystems. *Science*, **277**, July, 1997, 494--99.)
24. Various internet sources.